Julia Programming for Operations Research 2/e

Changhyun Kwon

Julia Programming for Operations Research

https://www.chkwon.net/julia

Second Edition

Published by Changhyun Kwon
Cover Design by Joo Yeon Woo / www.spacekite.net
Cat Drawing by Bomin Kwon

Copyright © 2019 by Changhyun Kwon
All Rights Reserved.
version 2021/03/06 13:58:00

Contents

1 **Introduction and Installation**   **1**
   1.1   What is Julia and Why Julia? 2
   1.2   Installing Julia 4
       1.2.1   Installing Julia in Windows 4
       1.2.2   Installing Julia in macOS 10
       1.2.3   Running Julia Scripts 15
       1.2.4   Installing Gurobi 15
       1.2.5   Installing CPLEX 16
   1.3   Installing IJulia 18
   1.4   Package Management 21
   1.5   Help 25

2 **Simple Linear Optimization**   **29**
   2.1   Linear Programming (LP) Problems 30
   2.2   Alternative Ways of Writing LP Problems 34
   2.3   Yet Another Way of Writing LP Problems 37
   2.4   Mixed Integer Linear Programming (MILP) Problems 38

3 Basics of the Julia Language — 41
- 3.1 Vector, Matrix, and Array — 41
- 3.2 Tuple — 47
- 3.3 Indices and Ranges — 48
- 3.4 Printing Messages — 51
- 3.5 Collection, Dictionary, and For-Loop — 54
- 3.6 Function — 57
- 3.7 Scope of Variables — 59
- 3.8 Random Number Generation — 63
- 3.9 File Input/Output — 67
- 3.10 Plotting — 72
  - 3.10.1 The `PyPlot` Package — 72
  - 3.10.2 Avoiding Type-3 Fonts in `PyPlot` — 77

4 Selected Topics in Numerical Methods — 79
- 4.1 Curve Fitting — 79
- 4.2 Numerical Differentiation — 84
- 4.3 Numerical Integration — 87
- 4.4 Automatic Differentiation — 91

5 The Simplex Method — 95
- 5.1 A Brief Description of the Simplex Method — 95
- 5.2 Searching All Basic Feasible Solutions — 98
- 5.3 Using the JuMP Package — 104
- 5.4 Pivoting in Tableau Form — 105
- 5.5 Implementing the Simplex Method — 107
  - 5.5.1 `initialize(c, A, b)` — 109
  - 5.5.2 `is_optimal(tableau)` — 111
  - 5.5.3 `pivoting!(tableau)` — 112
  - 5.5.4 Creating a Module — 116
- 5.6 Next Steps — 122

6 Network Optimization Problems — 123
- 6.1 The Minimal-Cost Network-Flow Problem — 123
- 6.2 The Transportation Problem — 133
- 6.3 The Shortest Path Problem — 139

	6.4 Implementing Dijkstra's Algorithm	144

7 Interior Point Methods — 151
- 7.1 The Affine Scaling Algorithm 151
- 7.2 The Primal Path Following Algorithm 157
- 7.3 Remarks 162

8 Nonlinear Optimization Problems — 165
- 8.1 Unconstrained Optimization 165
  - 8.1.1 Line Search 165
  - 8.1.2 Unconstrained Optimization 167
  - 8.1.3 Box-constrained Optimization 168
- 8.2 Nonlinear Optimization 169
- 8.3 Other Solvers 170
- 8.4 Mixed Integer Nonlinear Programming 175

9 Monte Carlo Methods — 177
- 9.1 Probability Distributions 177
- 9.2 Randomized Linear Program 179
- 9.3 Estimating the Number of Simple Paths 186

10 Lagrangian Relaxation — 197
- 10.1 Introduction 197
  - 10.1.1 Lower and Upper Bounds 198
  - 10.1.2 Subgradient Optimization 200
  - 10.1.3 Summary 200
- 10.2 The p-Median Problem 201
  - 10.2.1 Reading the Data File 202
  - 10.2.2 Solving the p-Median Problem Optimally 204
  - 10.2.3 Lagrangian Relaxation 205
  - 10.2.4 Finding Lower Bounds 206
  - 10.2.5 Finding Upper Bounds 210
  - 10.2.6 Updating the Lagrangian Multiplier 212

11 Complementarity Problems — **225**
- 11.1 Linear Complementarity Problems (LCP) 225
- 11.2 Nonlinear Complementarity Problems (NCP) 233
- 11.3 Mixed Complementarity Problems (MCP) 237

12 Parameters in Optimization Solvers — **239**
- 12.1 Setting CPU Time Limit 239
- 12.2 Setting the Optimality Gap Tolerance 240
- 12.3 Warmstart 241
- 12.4 Big-M and Integrality Tolerance 242
- 12.5 Turning off the Solver Output 244
- 12.6 Other Solver Parameters 244

Index — **247**

Preface

The main motivation of writing this book was to help myself. I am a professor in the field of operations research, and my daily activities involve building models of mathematical optimization, developing algorithms for solving the problems, implementing those algorithms using computer programming languages, experimenting with data, etc. Three languages are involved: human language, mathematical language, and computer language. My students and I need to go over three different languages. We need "translation" among the three languages.

When my students seek help on the tasks of "translation," I often provide them with my prior translation as an example or find online resources that may be helpful to them. If students have proper background with proper mathematical education, sufficient computer programming experience, and good understanding of how numerical computing works, students can learn easier and my daily tasks in research and education would go smoothly.

To my frustration, however, many graduate students in operations research take long time to learn how to "translate." This book is to help them and help me to help them.

I'm neither a computer scientist nor a software engineer. Therefore, this book does not teach the best translation. Instead, I'll try to teach how one can finish some common tasks necessary in research and development works arising in the field of operations research and management science. It will be just one translation, not the best for sure. But after reading this book, readers will certainly be able to get things done, one way or the other.

What this book teaches

This book is *neither* a textbook in numerical methods, a comprehensive introductory book to Julia programming, a textbook on numerical optimization, a complete manual of optimization solvers, *nor* an introductory book to computational science and engineering—it is a little bit of all.

This book will first teach how to install the Julia Language itself. This book teaches a little bit of syntax and standard libraries of Julia, a little bit of programming skills using Julia, a little bit of numerical methods, a little bit of optimization modeling, a little bit of Monte Carlo methods, a little bit of algorithms, and a little bit of optimization solvers.

This book by no means is complete and cannot serve as a standalone textbook for any of the above-mentioned topics. In my opinion, it is best to use this book along with other major textbooks or reference books in operations research and management science. This book assumes that readers are already familiar with topics in optimization theory and algorithms or are willing to learn by themselves from other references. Of course, I provide the best references of my knowledge to each topic.

After reading this book and some coding exercises, readers should be able to search and read many other technical documents available online. This book will just help the first step to computing in operations research and management science. This book is literally a primer on computing.

How this book can be used

This book will certainly help graduate students (and their advisors) for tasks in their research. First year graduate students may use this book as a *tutorial* that guides them to various optimization solvers and algorithms available. This book will also be a *companion* through their graduate study. While students take various courses during their graduate study, this book will be always a good starting point to learn how to solve certain optimization problems and implement algorithms they learned. Eventually, this book can be a helpful *reference* for their thesis research.

Advanced graduate students may use this book as a *reference*. For example, when they need to implement a Lagrangian relaxation method for their own problem, they can refer to a chapter in this book to see how I did it and learn how they may be able to do it.

It is also my hope that this book can be used for courses in operations research, analytics, linear programming, nonlinear programming, numerical optimization, network optimization, management science, and transportation engineering, as a *supplementary textbook*. If there is a short course with 1 or 2 credit hours for teaching numerical methods and computing tools in operations research and management science, this book can be *primary or secondary textbook*, depending on the instructor's main focus.

Notes to advanced programmers

If you are already familiar with computing and at least one computer programming language, I don't think this book will have much value for you. There are many resources available on the web, and you will be able to learn about the Julia Language and catch up with the state-of-the-art easily. If you want to learn and catch up even faster with much less troubles, this book can be helpful.

I had some experiences with MATLAB and Java before learning Julia. Learning Julia was not very difficult, but exciting and fun. I just needed a good "excuse" to learn and use Julia. Check what my excuse was in the first chapter.

Acknowledgment

I sincerely appreciate all the efforts from Julia developers. The Julia Language is a beautiful language that I love very much. It changed my daily computing life completely. I am thankful to the developers of the JuMP and other related packages. After JuMP, I no longer look for better modeling languages. I am also grateful to Joo Yeon Woo for the cover design and Bomin Kwon for the cat drawing.

Tampa, Florida
Changhyun Kwon

1
Introduction and Installation

This chapter will introduce what the Julia Language is and explain why I love it. More importantly, this chapter will teach you how to obtain Julia and install it in your machine. Well, at this moment, the most challenging task for using Julia in computing would probably be installing the language and other libraries and programs correctly in your own machine. I will go over every step with fine details with screenshots for both Windows and Mac machines. I assumed that Linux users can handle the installation process well enough without much help from this book by reading online manuals and googling. Perhaps the Mac section could be useful to Linux users.

All Julia codes in this book are shared as a git repository and are available at the book website: http://www.chkwon.net/julia. Codes are tested with

- Julia v1.3.0

- JuMP v0.21.2

- Optim v0.20.6

I will introduce what JuMP and Optim are gradually later in the book.

1.1 What is Julia and Why Julia?

The Julia Language is a young emerging language, whose primary target is technical computing. It is developed for making technical computing more fun and more efficient. There are many good things about the Julia Language from the perspective of computer scientists and software engineers; you can read about the language at the official website[1].

Here is a quote from the creators of Julia from their first official blog article "Why We Created Julia"[2]:

> "We want a language that's open source, with a liberal license. We want the speed of C with the dynamism of Ruby. We want a language that's homoiconic, with true macros like Lisp, but with obvious, familiar mathematical notation like Matlab. We want something as usable for general programming as Python, as easy for statistics as R, as natural for string processing as Perl, as powerful for linear algebra as Matlab, as good at gluing programs together as the shell. Something that is dirt simple to learn, yet keeps the most serious hackers happy. We want it interactive and we want it compiled.
>
> (Did we mention it should be as fast as C?)"

So this is how Julia was created, to serve all above greedy wishes.

Let me tell you my story. I used to be a Java developer for a few years before I joined a graduate school. My first computer codes for homework assignments and course projects were naturally written in Java; even before then, I used C for my homework assignments for computing when I was an undergraduate student. Later, in the graduate school, I started using MATLAB, mainly because my fellow graduate students in the lab were using MATLAB. I needed to learn from them, so I used MATLAB.

I liked MATLAB. Unlike in Java and C, I don't need to declare every single variable before I use it; I just use it in MATLAB. Arrays are not just arrays in the computer memory; arrays in MATLAB are just like vectors and matrices. Plotting computation results is easy. For modeling optimization problems, I used GAMS

---

[1] http://julialang.org
[2] http://julialang.org/blog/2012/02/why-we-created-julia

and connected with solvers like CPLEX. While the MATLAB-GAMS-CPLEX chain suited my purpose well, I wasn't that happy with the syntax of GAMS—I couldn't fully understand—and the slow speed of the interface between GAMS and MATLAB. While CPLEX provides complete connectivities with C, Java, and Python, it was very basic with MATLAB.

When I finished with my graduate degree, I seriously considered Python. It was—and still is—a very popular choice for many computational scientists. CPLEX also has a better support for Python than MATLAB. Unlike MATLAB, Python is a free and open source language. However, I didn't go with Python and decided to stick with MATLAB. I personally don't like 0 being the first index of arrays in C and Java. In Python, it is also 0. In MATLAB, it is 1. For example, if we have a vector like:

$$\mathbf{v} = \begin{bmatrix} 1 \\ 0 \\ 3 \\ -1 \end{bmatrix}$$

it may be written in MATLAB as:

```
v = [1; 0; 3; -1]
```

The first element of this vector should be accessible by v(1), not v(0). The i-th element must be v(i), not v(i-1). So I stayed with MATLAB.

Later in 2012, the Julia Language was introduced and it looked attractive to me, since at least the array index begins with 1. After some investigations, I still didn't move to Julia at that time. It was ugly in supporting optimization modeling and solvers. I kept using MATLAB.

In 2014, I came across several blog articles and tweets talking about Julia again. I gave it one more look. Then I found a package for modeling optimization problems in Julia, called JuMP—Julia for Mathematical Programming. After spending a few hours, I fell in love with JuMP and decided to go with Julia, well more with JuMP. Here is a part of my code for solving a network optimization problem:

```
@variable(m, 0<= x[links] <=1)

@objective(m, Min, sum(c[(i,j)] * x[(i,j)] for (i,j) in links) )
```

1.2. Installing Julia

```
for i=1:no_node
  @constraint(m, sum(x[(ii,j)] for (ii,j) in links if ii==i )
              - sum(x[(j,ii)] for (j,ii) in links if ii==i ) == b[i])
end

optimize!(m)
```

This is indeed a direct "translation" of the following mathematical language:

$$\min \sum_{(i,j) \in \mathcal{A}} c_{ij} x_{ij}$$

subject to

$$\sum_{(i,j) \in \mathcal{A}} x_{ij} - \sum_{(j,i) \in \mathcal{A}} x_{ji} = b_i \quad \forall i \in \mathcal{N}$$

$$0 \leq x_{ij} \leq 1 \quad \forall (i,j) \in \mathcal{A}$$

I think it is a very obvious translation. It is quite beautiful, isn't it?

CPLEX and its competitor Gurobi are also very smoothly connected with Julia via JuMP. Why should I hesitate? After several years of using Julia, I still love it—I even wrote a book.

1.2 Installing Julia

Graduate students and researchers are strongly recommended to install Julia in their local computers. In this guide, we will first install Julia and then install two optimization packages, JuMP and GLPK. JuMP stands for 'Julia for Mathematical Programming', which is a modeling language for optimization problems. GLPK is an open-source linear optimization solver that can solve both continuous and discrete linear programs. Windows users go to Section 1.2.1, and Mac users go to Section 1.2.2.

Chapter 1. Introduction and Installation

1.2.1 Installing Julia in Windows

- **Step 1.** Download Julia from the official website.[3] (Select an appropriate version: 32-bit or 64-bit. 64-bit recommended whenever possible.)

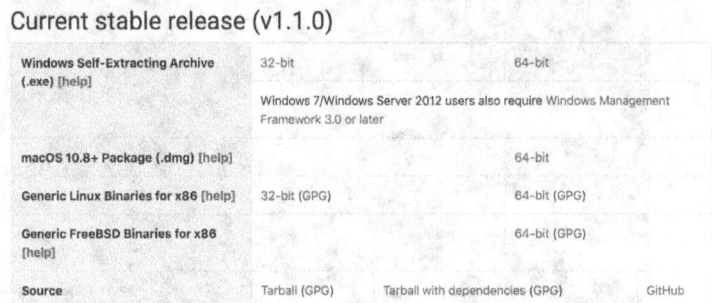

- **Step 2.** Install Julia in `C:\julia`. (You need to make the installation folder consistent with the path you set in Step 3.)

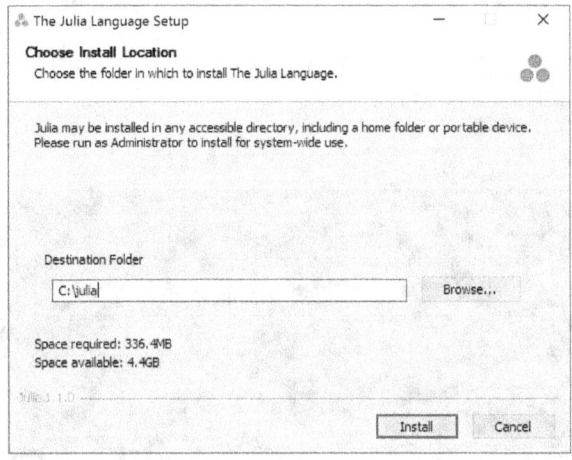

- **Step 3.** Open a Command Prompt and enter the following command:

---

[3]http://julialang.org/downloads/

1.2. Installing Julia

```
setx PATH "%PATH%;C:\julia\bin"
```

If you do not know how to open a Command Prompt, just google 'how to open command prompt windows.'

- **Step 4.** Open a **NEW** command prompt and type

```
echo %PATH%
```

The output must include `C:\julia\bin` in the end. If not, you must have something wrong.

Chapter 1. Introduction and Installation

- **Step 5.** Run julia.

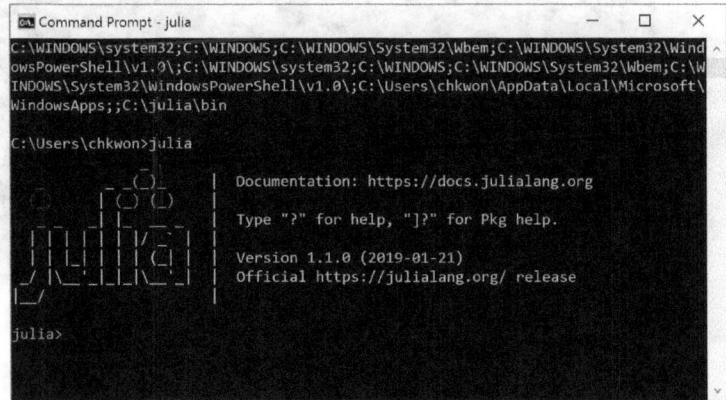

You have successfully installed the Julia Language on your Windows computer. Now it is time to install additional packages for mathematical optimization.

- **Step 6.** In your Julia prompt, type

```
julia> using Pkg
julia> Pkg.add("JuMP")
julia> Pkg.add("GLPK")
```

Installing the first package can take long time, because it initializes your Julia package folder and synchronizes with the entire package list.

1.2. Installing Julia

- **Step 7.** Open Notepad (or any other text editor such as Visual Studio Code[4]) and type the following, and save the file as `script.jl` in some folder of your choice.

```
using JuMP, GLPK
m = Model(GLPK.Optimizer)

@variable(m, 0 <= x <= 2 )
@variable(m, 0 <= y <= 30 )

@objective(m, Max, 5x + 3*y )

@constraint(m, 1x + 5y <= 3.0 )

JuMP.optimize!(m)
println("Objective value: ", JuMP.objective_value(m))
println("x = ", JuMP.value(x))
println("y = ", JuMP.value(y))
```

- **Step 8.** Press and hold your Shift Key and right-click the folder name, and choose "Open command window here."

---

[4] https://code.visualstudio.com

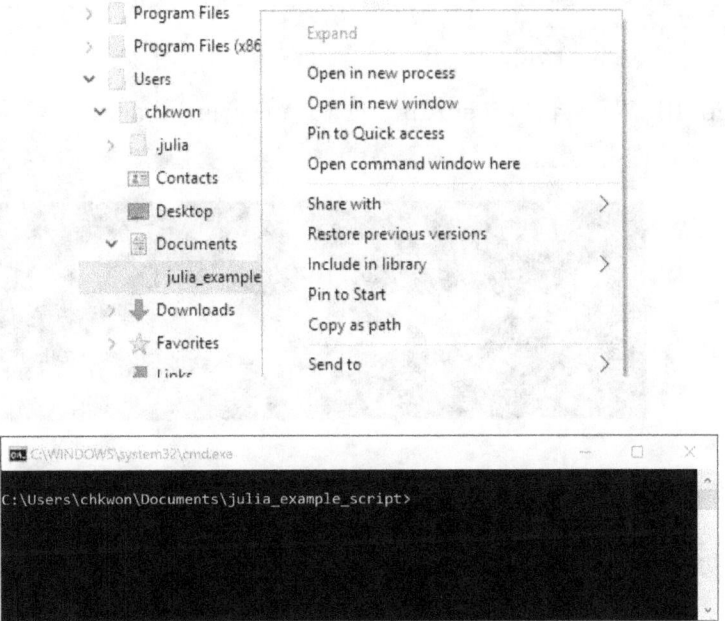

- **Step 9.** Type dir to see your script file script.jl.

If you see a filename such as script.jl.txt, use the following command to rename:

1.2. Installing Julia

```
ren script.jl.txt script.jl
```

- **Step 10.** Type `julia script.jl` to run your julia script.

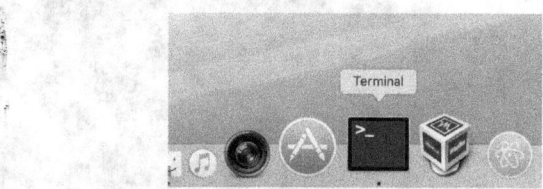

After a few seconds, the result of your julia script will be printed. Done.

Please proceed to Section 1.2.3.

1.2.2 Installing Julia in macOS

In macOS, we will use a package manager, called Homebrew. It provides a very convenient way of installing software in macOS.

- **Step 1.** Open "Terminal.app" from your Applications folder. (If you do not know how to open it, see this video.[5] It is convenient to place "Terminal.app" in your dock.

- **Step 2.** Visit `http://brew.sh` and follow the instruction to install Homebrew. It may ask you to enter your password to install Xcode Command Line Tools.

---
[5]`https://www.youtube.com/watch?v=zw7Nd67_aFw` "How to open the terminal window on a Mac"

Chapter 1. Introduction and Installation

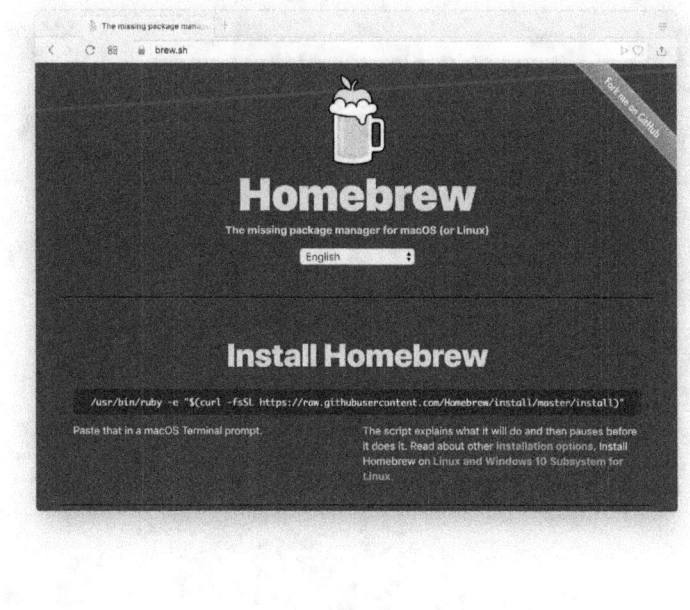

- **Step 3.** Installing Julia using Homebrew: In your terminal, enter the following command:

```
brew cask install julia
```

- **Step 5.** In your terminal, enter `julia`.

1.2. Installing Julia

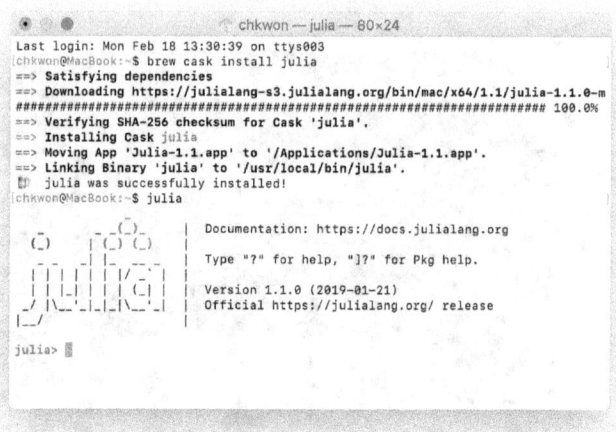

- **Step 6.** In your Julia prompt, type

```
julia> using Pkg
julia> Pkg.add("JuMP")
julia> Pkg.add("GLPK")
```

Installing the first package can take a long time, because it initializes your Julia package folder and synchronizes with the entire package list.

Chapter 1. Introduction and Installation

- **Step 7.** Open TextEdit (or any other text editor such as Visual Studio Code[6]) and type the following, and save the file as `script.jl` in some folder of your choice.

---
[6]https://code.visualstudio.com

1.2. Installing Julia

```
using JuMP, GLPK
m = Model(GLPK.Optimizer)

@variable(m, 0 <= x <= 2 )
@variable(m, 0 <= y <= 30 )

@objective(m, Max, 5x + 3*y )

@constraint(m, 1x + 5y <= 3.0 )

JuMP.optimize!(m)
println("Objective value: ", JuMP.objective_value(m))
println("x = ", JuMP.value(x))
println("y = ", JuMP.value(y))
```

- **Step 8.** Open a terminal window[7] at the folder that contains your `script.jl`.

- **Step 9.** Type `ls -al` to check your script file.

- **Step 10.** Type `julia script.jl` to run your script.

---

[7]To do this, you can drag the folder to the Terminal.app icon in your dock, or see http://osxdaily.com/2011/12/07/open-a-selected-finder-folder-in-a-new-terminal-window/

Chapter 1. Introduction and Installation

```
-rw-r--r--@ 1 chkwon  staff  310B Feb 20 22:50 script.jl
chkwon@MacBook:~/Documents/julia_example$ julia script.jl
Objective value: 10.6
x = 2.0
y = 0.2
chkwon@MacBook:~/Documents/julia_example$
```

After a few seconds, the result of your julia script will be printed. Done.

Please proceed to Section 1.2.3.

1.2.3 Running Julia Scripts

When you are ready, there are basically two methods to run your Julia script:

- In your Command Prompt or Terminal, enter `C:> julia your-script.jl`

- In your Julia prompt, enter `julia> include("your-script.jl")`.

1.2.4 Installing Gurobi

Instead of GLPK, one can use Gurobi, which is a commercial optimization solver package for solving LP, MILP, QP, MIQP, etc. Gurobi is free for students, teachers, professors, or anyone else related to educational organizations.

To install, follow these steps:

1. Download Gurobi Optimizer[8] and install in your computer. (You will need to register as an academic user.)

2. Request a free academic license[9] and follow their instructions to activate it.

---
[8] https://www.gurobi.com/downloads/gurobi-optimizer-eula/
[9] https://www.gurobi.com/academia/academic-program-and-licenses/

1.2. Installing Julia

3. Run Julia and add the `Gurobi` package. You need to tell Julia where Gurobi is installed:

   On Windows:

   ```
   julia> ENV["GUROBI_HOME"] =
           "C:\\Program Files\\gurobi910\\win64"
   julia> using Pkg
   julia> Pkg.add("Gurobi")
   ```

   On macOS:

   ```
   julia> ENV["GUROBI_HOME"] =
           "/Library/gurobi910/mac64"
   julia> using Pkg
   julia> Pkg.add("Gurobi")
   ```

4. Ready. Test the following code:

   ```
   using JuMP, Gurobi
   m = Model(Gurobi.Optimizer)
   @variable(m, x <= 5)
   @variable(m, y <= 45)
   @objective(m, Max, x + y)
   @constraint(m, 50x + 24y <= 2400)
   @constraint(m, 30x + 33y <= 2100)

   JuMP.optimize!(m)
   println("Objective value: ", JuMP.objective_value(m))
   println("x = ", JuMP.value(x))
   println("y = ", JuMP.value(y))
   ```

1.2.5 Installing CPLEX

Instead of Gurobi, you can install and connect the CPLEX solver, which is also free to academics.

You can follow this step by step guide to install:

Chapter 1. Introduction and Installation

1. Go to the IBM ILOG CPLEX Optimization Studio page[10].

2. Click 'Access free academic edition.'

3. Log in with your institution email and certify.

4. Download an appropriate version of IBM ILOG CPLEX Optimization Studio. It should be v12.10 or higher.

5. Run the downloaded file and install CPLEX. I recommend using the default installation folder.

6. Add the `CPLEX` package in Julia. You have to tell Julia where the CPLEX library is installed.

    On Windows:

    ```
    julia> ENV["CPLEX_STUDIO_BINARIES"] =
           "C:\\Program Files\\CPLEX_Studio1210\\cplex\\bin\\x86-64_win\\"
    julia> using Pkg
    julia> Pkg.add("CPLEX")
    julia> Pkg.build("CPLEX")
    ```

    On macOS:

    ```
    julia> ENV["CPLEX_STUDIO_BINARIES"] =
           "/Applications/CPLEX_Studio1210/cplex/bin/x86-64_osx/"
    julia> using Pkg
    julia> Pkg.add("CPLEX")
    julia> Pkg.build("CPLEX")
    ```

7. Ready. Test the following code:

    ```
    using JuMP, CPLEX
    m = Model(CPLEX.Optimizer)
    @variable(m, x <= 5)
    @variable(m, y <= 45)
    ```

---

[10] https://www.ibm.com/products/ilog-cplex-optimization-studio

```
@objective(m, Max, x + y)
@constraint(m, 50x + 24y <= 2400)
@constraint(m, 30x + 33y <= 2100)

JuMP.optimize!(m)
println("Objective value: ", JuMP.objective_value(m))
println("x = ", JuMP.value(x))
println("y = ", JuMP.value(y))
```

1.3 Installing IJulia

You can also use an interactive Julia environment in your local computer, called Jupyter Notebook[11]. Well, at first there was `IPython` notebook that was an interactive programming environment for the Python language. It has been popular, and now it is extended to cover many other languages such as R, Julia, Ruby, etc. The extension became the Jupyter Notebook project. For Julia, it is called `IJulia`, following the naming convention of `IPython`.

To use `IJulia`, we need a distribution of Python and Jupyter. Julia can automatically install a distribution for you, unless you want to install it by yourself. If you let Julia install Python and Jupyter, they will be private to Julia, i.e. you will not be able to use Python and Jupyter outside of Julia.

The following process will automatically install Python and Jupyter.

1. Open a new terminal window and run Julia. Initialize environment variables:

```
julia> ENV["PYTHON"] = ""
""

julia> ENV["JUPYTER"] = ""
""
```

2. Install `IJulia`:

---

[11] http://jupyter.org

Chapter 1. Introduction and Installation

```
julia> using Pkg
julia> Pkg.add("IJulia")
```

3. To open the IJulia notebook in your web browser:

```
julia> using IJulia
julia> notebook()
```

It will open a webpage in your browser that looks like the following screenshot:

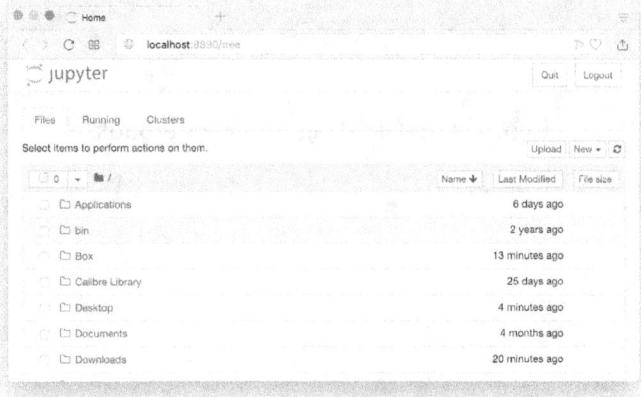

The current folder will be your home folder. You can move to another folder and also create a new folder by clicking the "New" button on the top-right corner of the screen. After locating a folder you want, you can now create a new IJulia notebook by clicking the "New" button again and select the julia version of yours, for example "Julia 1.1.0". See Figure 1.1.

It will basically open an interactive session of the Julia Language. If you have used Mathematica or Maple, the interface will look familiar. You can test basic Julia commands. When you need to evaluate a block of codes, press Shift+Enter, or press the "play" button. See Figure 1.2.

If you properly install a plotting package like PyPlot (details in Section 3.10.1), you can also do plotting directly within the IJulia notebook as shown in Figure 1.4.

19

1.3. Installing IJulia

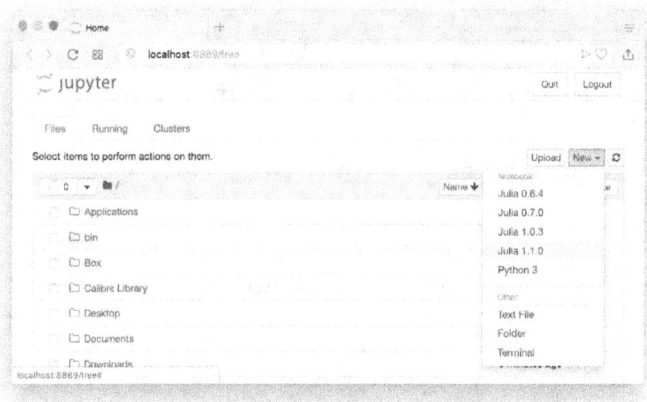

Figure 1.1: Creating a new notebook

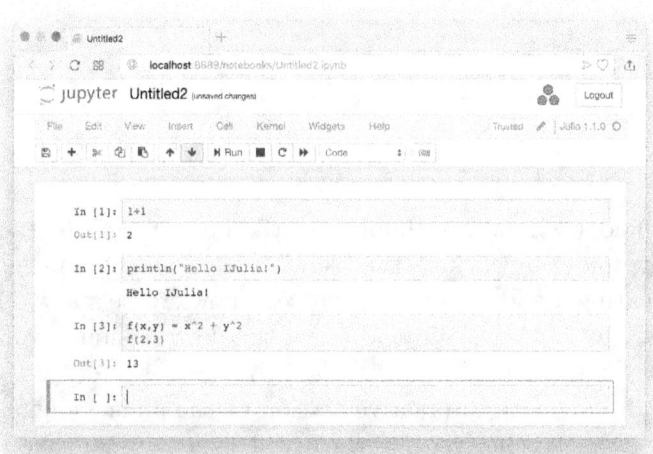

Figure 1.2: Some basic Julia codes.

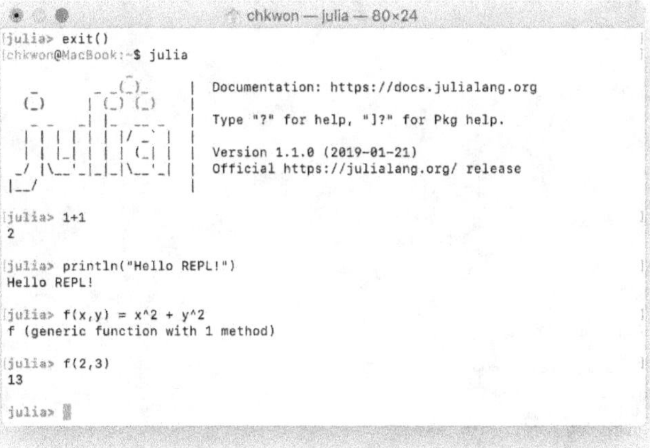

Figure 1.3: This is the REPL.

Personally, I prefer the REPL for most tasks, but I do occasionally use `IJulia`, especially when I need to test some simple things and need to plot the result quickly, or when I need to share the result of Julia computation with someone else. (`IJulia` can export the notebook in various formats, including HTML and PDF.)

What is REPL? It stands for read-eval-print loop. It is the Julia session that runs in your terminal; see Figure 1.3, which must look familiar to you already.

1.4 Package Management

There are many useful packages in Julia and we rely many parts of our computations on packages. If you have followed my instructions to install Julia, JuMP, Gurobi, and CPLEX, you have already installed a few packages. There are some more commands that are useful in managing packages.

```
julia> using Pkg
julia> Pkg.add("PackageName")
```

1.4. Package Management

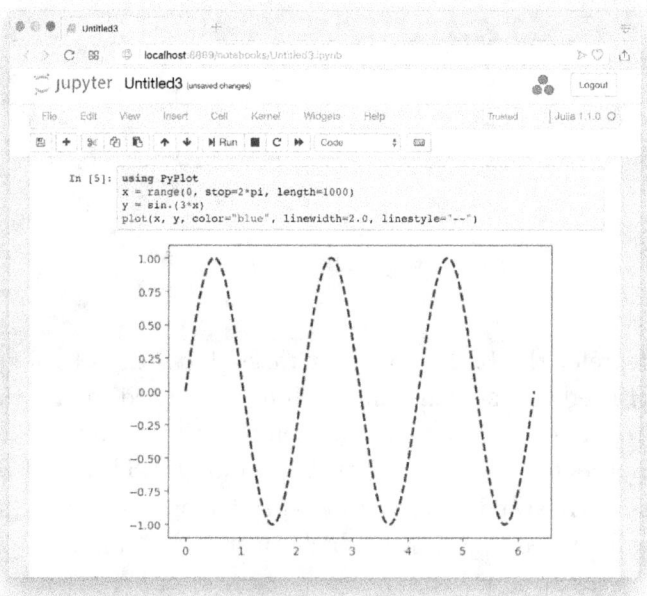

Figure 1.4: Plotting in IJulia

Figure 1.5: Package Mode in REPL

This installs a package, named `PackageName`. To find its online repository, you can just google the name `PackageName.jl`, and you will be directed to a repository hosted at `GitHub.com`.

Using `Pkg.add` requires `using Pkg` first. In REPL, by pressing the ']' key, you can enter the package management mode (Figure 1.5) and the prompt will change as follows:

```
(v1.3) pkg>
```

Then to install a package you can simply enter:

```
(v1.3) pkg> add PackageName
```

To install the JuMP package, you can do:

1.4. Package Management

```
(v1.3) pkg> add JuMP
```

To come back to the julia prompt, press the backspace or delete key.

```
julia> Pkg.rm("PackageName")
(v1.3) pkg> rm PackageName
```

This removes the package.

```
julia> Pkg.update()
(v1.3) pkg> update
```

This updates all packages that are already installed in your machine to the most recent versions.

```
julia> Pkg.status()
(v1.3) pkg> status
```

This displays what packages are installed and what their versions are. If you just want to know the version of a specific package, you can do:

```
julia> Pkg.installed()["PackageName"]
```

```
julia> Pkg.build("PackageName")
(v1.3) pkg> build PackageName
```

Occasionally, installing a package will fail during the `Pkg.add("PackageName")` process, usually because some libraries are not installed or system path variables are not configured correctly. Try to install some required libraries again and check the system path variables first. Then you may need to reboot your system or restart your Julia session. Then `Pkg.build("PackageName")`. Since you have downloaded package files during `Pkg.build("PackageName")`, you don't need to download them again; you just build it again.

1.5 Help

In REPL, you can use the Help mode. By pressing the ? key in REPL, you can enter the help mode. The prompt will change as follows:

```
help?>
```

Then type in any function name, for example, `println`, which results in:

```
help?> println
search: println printstyled print sprint isprint

  println([io::IO], xs...)

  Print (using print) xs followed by a newline. If io is not supplied, prints to
  stdout.

  Examples


  julia> println("Hello, world")
  Hello, world

  julia> io = IOBuffer();

  julia> println(io, "Hello, world")

  julia> String(take!(io))
  "Hello, world\n"
```

See also Figure 1.6.

Readers can find codes and other helpful resources in the author's website at

http://www.chkwon.net/julia

which also includes a link to a Facebook page of this book for discussion and communication.

This book does *not* teach everything of the Julia Language—only a very small part of it. When you want to learn more about the language, the first place you need to visit is

1.5. Help

Figure 1.6: Help Mode in REPL

http://julialang.org/learning/

where many helpful books, tutorials, videos, and articles are listed. Also, you will need to visit the official documentation of the Julia Language at

http://docs.julialang.org/

which I think serves as a good tutorial as well.

When you have a question, there will be many Julia enthusiasts ready for you. For questions and discussion, visit

https://discourse.julialang.org

and

http://julialang.org/community/

You can also ask questions at http://stackoverflow.com with tag julia-lang.

The webpage of JuMP is worth visiting for information about the JuMP.jl package.

http://jump.dev

1.5. Help

2 Simple Linear Optimization

This chapter provides a quick guide for solving simple linear optimization problems. For modeling, we use the JuMP package, and for computing, we use one of the following solvers.

- Clp: an open-source solver for linear programming (LP) problems from COIN-OR.

- Cbc: an open-source solver for mixed integer linear programming (MILP) problems from COIN-OR.

- GLPK: an open-source solver for mixed integer linear programming problem (MILP) problems from GNU.

- Gurobi: a commercial solver for both LP and MILP, free for academic users

- CPLEX: a commercial solver for both LP and MILP, free for academic users

Open-source solvers Clp, Cbc, and GLPK can be obtained by simply installing the corresponding Julia packages:

```
julia> using Pkg
julia> Pkg.add("Clp")
julia> Pkg.add("Cbc")
julia> Pkg.add("GLPK")
```

In fact, the `Clp` package automatically installs the `Cbc` package. COIN-OR is an open-source initiative, titled "Computational Infrastructure for Operations Research."

For commercial solvers Gurobi and CPLEX, one must first install the solver software, and then install the corresponding Julia packages:

```
julia> using Pkg
julia> Pkg.add("Gurobi")
julia> Pkg.add("CPLEX")
```

There are a couple of things to do before you add Julia packages. See Sections 1.2.4 and 1.2.5 for the details.

There are some alternatives available, both open-source and commercial solvers. See the list of available solvers via JuMP[1]. Nonlinear optimization solvers will be discussed in Chapter 8.

2.1 Linear Programming (LP) Problems

Once you have installed the JuMP package and an optimization solver mentioned above, we can have Julia solve linear programming (LP) and mixed integer linear programming (MILP) problems easily. For example, consider the following LP problem:

$$\max \quad x_1 + 2x_2 + 5x_3$$

subject to

$$-x_1 + x_2 + 3x_3 \leq -5$$
$$x_1 + 3x_2 - 7x_3 \leq 10$$
$$0 \leq x_1 \leq 10$$
$$x_2 \geq 0$$
$$x_3 \geq 0.$$

Using Julia and JuMP, we can write the following code:

---

[1] http://jump.dev/JuMP.jl/stable/installation/

Listing 2.1: LP Example 1
code/chap2/LP1.jl

```julia
using JuMP, GLPK

# Preparing an optimization model
m = Model(GLPK.Optimizer)

# Declaring variables
@variable(m, 0<= x1 <=10)
@variable(m, x2 >=0)
@variable(m, x3 >=0)

# Setting the objective
@objective(m, Max, x1 + 2x2 + 5x3)

# Adding constraints
@constraint(m, constraint1, -x1 +  x2 + 3x3 <= -5)
@constraint(m, constraint2,  x1 + 3x2 - 7x3 <= 10)

# Printing the prepared optimization model
print(m)

# Solving the optimization problem
JuMP.optimize!(m)

# Printing the optimal solutions obtained
println("Optimal Solutions:")
println("x1 = ", JuMP.value(x1))
println("x2 = ", JuMP.value(x2))
println("x3 = ", JuMP.value(x3))

# Printing the optimal dual variables
println("Dual Variables:")
println("dual1 = ", JuMP.shadow_price(constraint1))
println("dual2 = ", JuMP.shadow_price(constraint2))
```

The above code is pretty much self-explanatory, but here are some explanations. We first declare a placeholder for an optimization model:

2.1. Linear Programming (LP) Problems

```
m = Model(GLPK.Optimizer)
```

where we also indicated that we want to use the GLPK optimization solver. We call the model m.

We declare three variables:

```
@variable(m, 0<= x1 <=10)
@variable(m, x2 >= 0)
@variable(m, x3 >= 0)
```

where we used 'macros' from the JuMP package, @variable. In Julia, macros do repeated jobs for you. It is somewhat similar to 'functions' with some important differences. Refer to the official documentation[2].

Using another macro @objective, we set the objective:

```
@objective(m, Max, x1 + 2x2 + 5x3)
```

Two constraints are added by the @constraint macro:

```
@constraint(m, constraint1, -x1 +  x2 + 3x3 <= -5)
@constraint(m, constraint2,  x1 + 3x2 - 7x3 <= 10)
```

Note that constraint1 and constraint2 are the names of those constraints. These names will be useful for obtaining the corresponding dual variable values.

We are now ready with the optimization problem. If you like you can print the optimization model and check how it is written, the code is as simple as:

```
print(m)
```

We solve the optimization problem:

---

[2]http://docs.julialang.org/en/v1/manual/metaprogramming/#macros

Chapter 2. Simple Linear Optimization

```
JuMP.optimize!(m)
```

After solving the optimization problem, we can obtain the values of variables at the optimality by using the `JuMP.value()` function:

```
println("Optimal Solutions:")
println("x1 = ", JuMP.value(x1))
println("x2 = ", JuMP.value(x2))
println("x3 = ", JuMP.value(x3))
```

where `println()` is a function that puts some text in a line on the screen. If you don't want to change the line after you print the text, use the `print()` function instead.

To obtain the values of optimal dual variables, call `JuMP.shadow_price()` with the corresponding constraint names as follows:

```
println("Dual Variables:")
println("dual1 = ", JuMP.shadow_price(constraint1))
println("dual2 = ", JuMP.shadow_price(constraint2))
```

IMPORTANT: There is also the `JuMP.dual()` function defined. However, the sign of `JuMP.dual()` results might not be as you would expect, since it follows the convention of conic duality. For linear optimization problems, `JuMP.shadow_price()` provides dual variable values as defined in most standard textbooks. Please refer to the relevant discussion in the JuMP documentation[3].

In my machine, the output by Gurobi looks like:

```
julia> include("LP1.jl")
Max x1 + 2 x2 + 5 x3
Subject to
 x1   0.0
 x2   0.0
 x3   0.0
```

---

[3] http://jump.dev/JuMP.jl/stable/constraints/#constraint_duality-1

2.2. Alternative Ways of Writing LP Problems

```
 x1   10.0
-x1 + x2 + 3 x3   -5.0
 x1 + 3 x2 - 7 x3   10.0
Optimal Solutions:
x1 = 10.0
x2 = 2.1875
x3 = 0.9375
Dual Variables:
dual1 = 1.8125
dual2 = 0.06249999999999998
```

If you want to use the Gurobi optimization solver instead of GLPK, use the following inputs:

```
using JuMP, Gurobi
m = Model(Gurobi.Optimizer)
```

For CPLEX:

```
using JuMP, CPLEX
m = Model(CPLEX.Optimizer)
```

There are many other optimization solvers supported by the JuMP package. See the manual of JuMP for a list.[4]

2.2 Alternative Ways of Writing LP Problems

We can use arrays to define variables. For the same LP problem as in the previous section, we can write a Julia code alternatively as follows:

To define the variable **x** as a three-dimensional vector, we can write:

```
@variable(m, x[1:3] >= 0)
```

---

[4] http://jump.dev/JuMP.jl/stable/installation/

where 1:3 means an array with numbers from 1 to 3 (incrementing by 1).

Then we prepare a column vector **c** and use it for defining the objective function:

```
c = [1; 2; 5]
@objective(m, Max, sum( c[i]*x[i] for i in 1:3))
```

which is essentially same as:

$$\max \sum_{i=1}^{3} c_i x_i$$

In LP problems, constraints are usually written in the vector-matrix notation as **Ax ≤ b**. Following this convention, we prepare a matrix **A** and a vector **b**, and use them for adding constraints:

```
A = [-1  1  3;
      1  3 -7]
b = [-5; 10]
@constraint(m, constraint1, sum( A[1,i]*x[i] for i in 1:3) <= b[1] )
@constraint(m, constraint2, sum( A[2,i]*x[i] for i in 1:3) <= b[2] )
```

This will be impractical, if we have 100 constraints, instead of just 2. Alternatively, we can write:

```
constraint = Dict()
for j in 1:2
   constraint[j] = @constraint(m, sum(A[j,i]*x[i] for i in 1:3) <= b[j])
end
```

Even better, we can also write:

```
@constraint(m, constraint[j in 1:2], sum(A[j,i]*x[i] for i in 1:3) <= b[j])
```

Use any form that works for you. The JuMP package provides many different methods of adding constraints. Read the official document.[5]

Finally, we add the bound constraint on x_1:

---

[5] http://jump.dev/JuMP.jl/stable/constraints/

2.2. Alternative Ways of Writing LP Problems

```
@constraint(m, bound, x[1] <= 10)
```

The final code is presented:

Listing 2.2: LP Example 2
code/chap2/LP2.jl

```julia
using JuMP, GLPK
m = Model(GLPK.Optimizer)

c = [ 1; 2; 5]
A = [-1  1  3;
      1  3 -7]
b = [-5; 10]

@variable(m, x[1:3] >= 0)
@objective(m, Max, sum( c[i]*x[i] for i in 1:3 ) )

@constraint(m, constraint[j in 1:2], sum( A[j,i]*x[i] for i in 1:3 ) <= b[j] )
@constraint(m, bound, x[1] <= 10)

JuMP.optimize!(m)


println("Optimal Solutions:")
for i in 1:3
  println("x[$i] = ", JuMP.value(x[i]))
end

println("Dual Variables:")
for j in 1:2
  println("dual[$j] = ", JuMP.shadow_price(constraint[j]))
end
```

Note that there have been changes in the code for printing. The result looks like:

```
julia> include("LP2.jl")
Optimal Solutions:
x[1] = 10.0
```

```
x[2] = 2.1875
x[3] = 0.9375
Dual Variables:
dual[1] = 1.8125
dual[2] = 0.06250000000000003
```

2.3 Yet Another Way of Writing LP Problems

In LP2.jl, we used 1:3 for indices for x_i and 1:2 for indices of constraints. If you want to change the data and solve another problem with the same structure, then you will have to change the numbers 3 and 2 manually, which of course is very tedious and will most likely create unwanted bugs. Instead, we can assign names for those sets of indices:

```
index_x = 1:3
index_constraints = 1:2
```

Then, you can rewrite the code for adding constraints, for example:

```
@constraint(m, constraint[j in index_constraints],
            sum( A[j,i]*x[i] for i in index_x ) <= b[j] )
```

The complete code would look like:

Listing 2.3: LP Example 3
code/chap2/LP3.jl

```
using JuMP, GLPK
m = Model(GLPK.Optimizer)

c = [ 1; 2; 5]
A = [-1  1  3;
      1  3 -7]
b = [-5; 10]

index_x = 1:3
```

2.4. Mixed Integer Linear Programming (MILP) Problems

```
index_constraints = 1:2

@variable(m, x[index_x] >= 0)
@objective(m, Max, sum( c[i]*x[i] for i in index_x) )

@constraint(m, constraint[j in index_constraints],
            sum( A[j,i]*x[i] for i in index_x ) <= b[j] )

@constraint(m, bound, x[1] <= 10)

JuMP.optimize!(m)

println("Optimal Solutions:")
for i in index_x
  println("x[$i] = ", JuMP.value(x[i]))
end

println("Dual Variables:")
for j in index_constraints
  println("dual[$j] = ", JuMP.shadow_price(constraint[j]))
end
```

The result of LP3.jl should be same of that of LP2.jl.

2.4 Mixed Integer Linear Programming (MILP) Problems

In many applications, variables are often binary or discrete; the resulting optimization problem then becomes an integer programming problem. Further if everything is linear and there are both continuous variables and integer variables, the optimization problem is called a mixed integer linear programming (MILP) problem. The Gurobi and CPLEX optimization solvers can handle this type of problem very well.

Suppose now x_2 is an integer variable and x_3 is a binary variable in the previous LP problem. That is:

$$\max \quad x_1 + 2x_2 + 5x_3$$

subject to

$$-x_1 + x_2 + 3x_3 \leq -5$$

Chapter 2. Simple Linear Optimization

$$x_1 + 3x_2 - 7x_3 \leq 10$$
$$0 \leq x_1 \leq 10$$
$$x_2 \geq 0 \text{ Integer}$$
$$x_3 \in \{0, 1\}.$$

Using JuMP, it is very simple to specify integer and binary variables. We can define variables as follows:

```julia
@variable(m, 0<= x1 <=10)
@variable(m, x2 >=0, Int)
@variable(m, x3, Bin)
```

The complete code would look like:

Listing 2.4: MILP Example 1
code/chap2/MILP1.jl

```julia
using JuMP, GLPK

# Preparing an optimization model
m = Model(GLPK.Optimizer)

# Declaring variables
@variable(m, 0<= x1 <=10)
@variable(m, x2 >=0, Int)
@variable(m, x3, Bin)

# Setting the objective
@objective(m, Max, x1 + 2x2 + 5x3)

# Adding constraints
@constraint(m, constraint1, -x1 +  x2 + 3x3 <= -5)
@constraint(m, constraint2,  x1 + 3x2 - 7x3 <= 10)

# Printing the prepared optimization model
print(m)

# Solving the optimization problem
JuMP.optimize!(m)
```

2.4. Mixed Integer Linear Programming (MILP) Problems

```julia
# Printing the optimal solutions obtained
println("Optimal Solutions:")
println("x1 = ", JuMP.value(x1))
println("x2 = ", JuMP.value(x2))
println("x3 = ", JuMP.value(x3))
```

The result looks like:

```
julia> include("MILP1.jl")
Max x1 + 2 x2 + 5 x3
Subject to
 x3 binary
 x2 integer
 x1  0.0
 x2  0.0
 x1  10.0
 -x1 + x2 + 3 x3  -5.0
 x1 + 3 x2 - 7 x3  10.0
Optimal Solutions:
x1 = 10.0
x2 = 2.0
x3 = 1.0
```

Basics of the Julia Language

In this chapter, I cover how we can do most common tasks for computing in operations research and management science with the Julia Language. While I will cover some part of the syntax of Julia, readers must consult with the official documentation[1] of Julia for other unexplained usages.

3.1 Vector, Matrix, and Array

Like MATLAB and many other computer languages for numerical computation, Julia provides easy and convenient, but strong, ways of handling vectors and matrices. For example, if you want to create vectors and matrices like

$$\mathbf{a} = \begin{bmatrix} 1 \\ 2 \\ 3 \end{bmatrix}, \qquad \mathbf{b} = \begin{bmatrix} 4 & 5 & 6 \end{bmatrix}, \qquad \mathbf{A} = \begin{bmatrix} 1 & 2 & 3 \\ 4 & 5 & 6 \end{bmatrix}$$

then in Julia, you can simply type

```
a = [1; 2; 3]
b = [4 5 6]
A = [1 2 3; 4 5 6]
```

---

[1] http://docs.julialang.org/

3.1. Vector, Matrix, and Array

where the semicolon (;) means a new row. Julia will return:

```
julia> a = [1; 2; 3]
3-element Array{Int64,1}:
 1
 2
 3

julia> b = [4 5 6]
1x3 Array{Int64,2}:
 4  5  6

julia> A = [1 2 3; 4 5 6]
2x3 Array{Int64,2}:
 1  2  3
 4  5  6
```

We can access the (i,j)-element of **A** by `A[i,j]`:

```
julia> A[1,3]
3

julia> A[2,1]
4
```

The transpose of vectors and matrices is easily obtained either of the following codes:

```
julia> transpose(A)
3x2 Array{Int64,2}:
 1  4
 2  5
 3  6

julia> A'
3x2 Array{Int64,2}:
 1  4
 2  5
 3  6
```

Let us introduce two column vectors:

```
a = [1; 2; 3]
c = [7; 8; 9]
```

The inner product, or dot product, may be obtained by the following way:

```
julia> a'*c
50
```

Another way is to use the `dot()` function. This function is provided in a standard library of Julia, called the `LinearAlgebra` package.

```
julia> using LinearAlgebra
julia> dot(a,c)
50
```

For many other useful functions in the `LinearAlgebra` package, see the official document[2].

The identity matrices of certain sizes:

```
julia> Matrix(1.0I, 2, 2)
2x2 Array{Float64,2}:
 1.0  0.0
 0.0  1.0

julia> Matrix(1.0I, 3, 3)
3x3 Array{Float64,2}:
 1.0  0.0  0.0
 0.0  1.0  0.0
 0.0  0.0  1.0
```

The matrices of zeros and ones of custom sizes:

---

[2]https://docs.julialang.org/en/v1/stdlib/LinearAlgebra/index.html

3.1. Vector, Matrix, and Array

```
julia> zeros(4,1)
4x1 Array{Float64,2}:
 0.0
 0.0
 0.0
 0.0

julia> zeros(2,3)
2x3 Array{Float64,2}:
 0.0  0.0  0.0
 0.0  0.0  0.0

julia> ones(1,3)
1x3 Array{Float64,2}:
 1.0  1.0  1.0

julia> ones(3,2)
3x2 Array{Float64,2}:
 1.0  1.0
 1.0  1.0
 1.0  1.0
```

When we have a square matrix

```
julia> B = [1 3 2; 3 2 2; 1 1 1]
3x3 Array{Int64,2}:
 1  3  2
 3  2  2
 1  1  1
```

its inverse can be computed:

```
julia> inv(B)
3x3 Array{Float64,2}:
  1.11022e-16   1.0  -2.0
  1.0           1.0  -4.0
 -1.0          -2.0   7.0
```

Of course, there are some numerical errors:

```
julia> B * inv(B)
3x3 Array{Float64,2}:
  1.0          0.0  0.0
  0.0          1.0  0.0
 -2.22045e-16  0.0  1.0
```

Note that the off-diagonal elements are not exactly zero. This is because the computation of the inverse matrix is not exact. For example, the (2,1)-element of the inverse matrix is not exactly 1, but:

```
julia> inv(B)[2,1]
1.0000000000000004
```

In the above, we have seen something like `Int64` and `Float64`. In 32-bit systems, it would have been `Int32` and `Float32`. These are data types. If the elements in your vectors and matrices are integers for sure, you can use `Int64`. On the other hand, if any element is non-integer values, such as 1.0000000000000004, you need to use `Float64`. These are usually done automatically:

```
julia> a = [1; 2; 3]
3-element Array{Int64,1}:
 1
 2
 3

julia> b = [1.0; 2; 3]
3-element Array{Float64,1}:
 1.0
 2.0
 3.0
```

In some cases, you will want to first create an array object of a certain type, then assign values. This can be done by calling `Array` with a keyword `undef`. For example, if we want an array of `Float64` data type and of size 3, then we can do:

3.1. Vector, Matrix, and Array

```
julia> d = Array{Float64}(undef, 3)
3-element Array{Float64,1}:
 2.287669227e-314
 2.2121401076e-314
 2.2147848607e-314
```

Some values that are close to zero are pre-assigned. Now you can assign the value you want:

```
julia> d[1] = 1
1

julia> d[2] = 2
2

julia> d[3] = 3
3

julia> d
3-element Array{Float64,1}:
 1.0
 2.0
 3.0
```

Although you entered the exact integer values, the resulting array is of `Float64` type, as it was originally defined.

If your array represents a vector or a matrix, I recommend you create an array by explicitly specifying the dimension. For a 3-by-1 column vector, you have to do:

```
p = Array{Float64}(undef, 3, 1)
```

and for a 1-by-3 row vector, you have to do:

```
q = Array{Float64}(undef, 1, 3)
```

Then the products can be written as `p*q` or `q*p`.

For more information on functions related to arrays, please refer to the official document.[3]

3.2 Tuple

The data type of arrays can be a pair of data types. For example, if we want to store (1,2), (2,3), and (3,4) in an array, we can create an array of type (`Int64`, `Int64`).

```
julia> pairs = Array{Tuple{Int64, Int64}}(undef, 3)
3-element Array{Tuple{Int64,Int64},1}:
 (4475364336, 4639719056)
 (4616151040, 4596384320)
 (0, 4596384320)

julia> pairs[1] = (1,2)
(1, 2)

julia> pairs[2] = (2,3)
(2, 3)

julia> pairs[3] = (3,4)
(3, 4)

julia> pairs
3-element Array{Tuple{Int64,Int64},1}:
 (1,2)
 (2,3)
 (3,4)
```

This is same as:

```
julia> pairs = [ (1,2); (2,3); (3,4) ]
3-element Array{(Int64,Int64),1}:
 (1, 2)
 (2, 3)
 (3, 4)
```

---

[3]https://docs.julialang.org/en/v1/manual/arrays/

3.3. Indices and Ranges

These types of arrays can be useful for handling network data with nodes and links, for example for storing data like (i, j). When you want (i, j, k), simply do:

```
julia> ijk_array = Array{Tuple{Int64, Int64, Int64}}(undef, 3)
3-element Array{Tuple{Int64,Int64,Int64},1}:
 (4634730720, 4634730640, 1)
 (4634730800, 4634730880, 4634730960)
 (3, 2, 2)

julia> ijk_array[1] = (1,4,2)
(1, 4, 2)

...
```

3.3 Indices and Ranges

When we are dealing with indices of arrays—vectors, matrices, or any other arrays—a range will be useful. If we want a set of indices from 1 to 9, we can simply do `1:9`. If we want steps of 2, we do `1:2:9`. Suppose we have a vector **a**:

```
julia> a = [10; 20; 30; 40; 50; 60; 70; 80; 90]
9-element Array{Int64,1}:
 10
 20
 30
 40
 50
 60
 70
 80
 90
```

If we want the first three elements, we can do:

```
julia> a[1:3]
3-element Array{Int64,1}:
 10
 20
 30
```

Some other examples:

```
julia> a[1:3:9]
3-element Array{Int64,1}:
 10
 40
 70
```

where `1:3:9` refers to 1, 4, and 7. Although we specified 9 in `1:3:9`, it didn't reach 9.

The last index can be accessed by using a special keyword **end**. The last three elements can be accessed by:

```
julia> a[end-2:end]
3-element Array{Int64,1}:
 70
 80
 90
```

We can also assign some values using a range:

```
julia> b = [200; 300; 400]
3-element Array{Int64,1}:
 200
 300
 400

julia> a[2:4] = b
3-element Array{Int64,1}:
 200
 300
 400

julia> a
9-element Array{Int64,1}:
  10
 200
 300
 400
```

3.3. Indices and Ranges

```
50
60
70
80
90
```

We can also use ranges to define an array. Just toss a range to the `collect()` function:

```
julia> c = collect(1:2:9)
5-element Array{Int64,1}:
 1
 3
 5
 7
 9
```

Suppose we have a matrix **A**:

```
julia> A= [1 2 3; 4 5 6; 7 8 9]
3x3 Array{Int64,2}:
 1 2 3
 4 5 6
 7 8 9
```

The second column of **A** can be accessed by:

```
julia> A[:, 2]
3-element Array{Int64,1}:
 2
 5
 8
```

The second to third columns of **A**:

```
julia> A[:, 2:3]
3×2 Array{Int64,2}:
 2  3
 5  6
 8  9
```

The third row can be accessed by:

```
julia> A[3, :]
3-element Array{Int64,1}:
 7
 8
 9
```

Note that it returns a column vector. To obtain a row vector, one may consider:

```
julia> A[3:3, :]
1×3 Array{Int64,2}:
 7  8  9
```

The second to third rows can be obtained by:

```
julia> A[2:3, :]
2×3 Array{Int64,2}:
 4  5  6
 7  8  9
```

3.4 Printing Messages

Displaying messages on the screen is the simplest tool for debugging and checking the computational results. This section introduces functions related to displaying and printing messages.

The most frequently used printing function would be `println()` and `print()`. We can simply do:

3.4. Printing Messages

```
julia> println("Hello World")
Hello World
```

The difference between the two functions is that `println()` adds an empty line after it.

```
julia> print("Hello "); print("World"); print(" Again")
Hello World Again
julia> println("Hello "); println("World"); println(" Again")
Hello
World
 Again

julia>
```

Combining the custom text with values in a variable is easy:

```
julia> a = 123.0
123.0

julia> println("The value of a = ", a)
The value of a = 123.0

julia> println("a is $a, and a-10 is $(a-10).")
a is 123.0, and a-10 is 113.0.
```

It works for arrays:

```
julia> b = [1; 3; 10]
3-element Array{Int64,1}:
  1
  3
 10

julia> println("b is $b.")
b is [1,3,10].

julia> println("The second element of b is $(b[2]).")
The second element of b is 3.
```

More advanced functionalities are provided by the @printf macro, which uses the style of printf() of C. This macro is provided by the Printf package. Here is an example:

```
julia> using Printf
julia> @printf("The %s of a = %f", "value", a)
The value of a = 123.000000
```

The first argument is the format specification: it involves a string with symbols like %s and %f. In the format specification, %s is a placeholder for a string like "value", and %f is a placeholder for a floating number like a. The placeholder for an integer is %d.

The @printf macro is very useful when we want to print a series of numbers whose digit number differ, but we want to give some good alignment options. Suppose

```
c = [ 123.12345    ;
       10.983      ;
        1.0932132 ]
```

If we just use println(), we obtain:

```
julia> for i in 1:length(c)
           println("c[$i] = $(c[i])")
       end
c[1] = 123.12345
c[2] = 10.983
c[3] = 1.0932132
```

Not so pretty. Use @printf:

```
julia> for i in 1:length(c)
           @printf("c[%d] = %7.3f\n", i, c[i])
       end
c[1] = 123.123
c[2] =  10.983
c[3] =   1.093
```

3.5. Collection, Dictionary, and For-Loop

where `%7.3` indicates that we want the total number of digits to be 7 (including the decimal point) and the number of digits below the decimal point to be 3. In addition, `\n` means a new line.

There is another macro called `@sprintf`, which is basically the same as `@printf`, but it returns a string, instead of printing it on the screen. For example:

```
julia> str = @sprintf("The %s of a = %f", "value", a)
julia> println(str)
The value of a = 123.000000
```

For more format specifiers, see http://www.cplusplus.com/reference/cstdio/printf/.

3.5 Collection, Dictionary, and For-Loop

When we repeat a similar job for multiple times, we use a for-loop. For example:

```
julia> for i in 1:5
           println("This is number $i.")
       end
This is number 1.
This is number 2.
This is number 3.
This is number 4.
This is number 5.
```

A for-loop has a structure of

```
for i in I
    # do something here for each i
end
```

where `I` is a *collection*. The most common collection type in computing is a range, like `1:5` in the above example.

If you want to stop at a certain point, you can **break** the loop:

Chapter 3. Basics of the Julia Language

```
julia> for i in 1:5
           if i >= 3
               break
           end
           println("This is number $i.")
       end
This is number 1.
This is number 2.
```

Another very useful collection type is *Dictionary*. A dictionary has keys and values. For example, suppose we have the following keys and values:

```
my_keys = ["Zinedine Zidane", "Magic Johnson", "Yuna Kim"]
my_values = ["football", "basketball", "figure skating"]
```

We can create a dictionary:

```
julia> d = Dict()
Dict{Any,Any} with 0 entries

julia> for i in 1:length(my_keys)
           d[my_keys[i]] = my_values[i]
       end

julia> d
Dict{Any,Any} with 3 entries:
  "Magic Johnson"   => "basketball"
  "Zinedine Zidane" => "football"
  "Yuna Kim"        => "figure skating"
```

When we use a dictionary, the order of elements saved in the dictionary should not be important. Note in the dictionary d, the order is not same as the order in k and v. If the order is important, we have to be careful with dictionaries.

Using the dictionary d defined above, we can, for example, do something like:

3.5. Collection, Dictionary, and For-Loop

```
julia> for (key, value) in d
           println("$key is a $value player.")
       end
Magic Johnson is a basketball player.
Zinedine Zidane is a football player.
Yuna Kim is a figure skating player.
```

We can also add a new element by:

```
julia> d["Diego Maradona"] = "football"
"football"

julia> d
Dict{Any,Any} with 4 entries:
  "Magic Johnson"    => "basketball"
  "Zinedine Zidane"  => "football"
  "Diego Maradona"   => "football"
  "Yuna Kim"         => "figure skating"
```

For a network, suppose we have the following data:

```
links = [ (1,2), (3,4), (4,2) ]
link_costs = [ 5, 13, 8 ]
```

We can create a dictionary for this data:

```
julia> link_dict = Dict()
Dict{Any,Any} with 0 entries

julia> for i in 1:length(links)
           link_dict[ links[i] ] = link_costs[i]
       end

julia> link_dict
Dict{Any,Any} with 3 entries:
  (1,2) => 5
  (4,2) => 8
  (3,4) => 13
```

Then we can use it as follows:

```
julia> for (link, cost) in link_dict
           println("Link $link has cost of $cost.")
       end
Link (1,2) has cost of 5.
Link (4,2) has cost of 8.
Link (3,4) has cost of 13.
```

For more information on collection and dictionary, please refer to the official document[4]. There are many convenient functions available.

Sometimes `while`-loops are more useful than `for`-loops. For the usage of `while`-loops and other flow controls, see the official document[5].

3.6 Function

Just like any mathematical function, functions in Julia accept inputs and return outputs. Consider a simple mathematical function:

$$f(x, y) = 3x + y$$

We can create a Julia function for this as follows:

```
function f(x,y)
   return 3x + y
end
```

Simple. Call it:

```
julia> f(1,3)
6

julia> 3 * ( f(3,2) + f(5,6) )
96
```

---

[4]https://docs.julialang.org/en/v1/base/collections/
[5]https://docs.julialang.org/en/v1/manual/control-flow/

3.6. Function

Alternatively, you can define the same function in more compact form, called "assignment form":

```
julia> f(x,y) = 3x+y
f (generic function with 1 method)

julia> f(1,3)
6
```

Functions can have multiple return values. For example:

```
function my_func(n, m)
  a = zeros(n,1)
  b = ones(m,1)
  return a, b
end
```

We can receive return values as follows:

```
julia> x, y = my_func(3,2)
([0.0; 0.0; 0.0], [1.0; 1.0])

julia> x
3x1 Array{Float64,2}:
 0.0
 0.0
 0.0

julia> y
2x1 Array{Float64,2}:
 1.0
 1.0
```

When any function `f` is defined for scalar quantities, we can use `f.` for vectorized operations. For example, consider `sqrt()`, which is defined for a scalar quantity:

```
julia> sqrt(9)
3.0

julia> sqrt([9 16])
ERROR: DimensionMismatch("matrix is not square: dimensions are (1, 2)")
```

For a vector quantity, we can use **sqrt.()** as follows:

```
julia> sqrt.([9 16])
1⨯2 Array{Float64,2}:
 3.0  4.0
```

Note that the square-root operation is applied to each element.

Adding a dot to the end of the function name can be applied any function.

```
julia> myfunc(x) = sin(x) + 3*x
myfunc (generic function with 1 method)

julia> myfunc(3)
9.141120008059866

julia> myfunc([5 10])
ERROR: DimensionMismatch("matrix is not square: dimensions are (1, 2)")

julia> myfunc.([5 10])
1⨯2 Array{Float64,2}:
 14.0411  29.456
```

More details about functions are found in the official document[6].

3.7 Scope of Variables

In any programming language, it is important to understand what variables can be accessed where. Let's consider the following code:

---

[6] https://docs.julialang.org/en/v1/manual/functions/

3.7. Scope of Variables

```
function f(x)
    return x+2
end

function g(x)
    return 3x+3
end
```

In this code, although the same variable name x is used in two different functions, the two x variables do not conflict. It is because they are defined in different *scope blocks*. Examples of scope blocks are `function` bodies, `for` loops, and `while` loops.

When a variable is defined or first introduced, the variable is accessible within the scope block and its sub scope blocks. As an example, consider a script file with the following code:

```
function f(x)
    return x+a
end

function run()
    a = 10
    return f(5)
end

run()
```

This will generate an error. To fix the problem, we should define or introduce first the a variable in the *parent* scope block, which is outside the function blocks. A revised code is:

```
function f(x)
    return x+a
end

function run()
    return f(5)
end
```

```
a = 10
run()
```

which returns:

```
15
```

Let's consider another example.

```
function f2(x)
  a = 0
  return x+a
end

a = 5
println(f2(1))
println(a)
```

What will be the printed values of f2(1) and a? Quite confusing. If you can avoid codes like the above example, I think it will be the best. The same `a` variable name is used in two different places, and it can be quite confusing sometimes. It may lead to serious bugs in your codes. The result of the above example is:

```
1
5
```

To better control the scope of variables, we can use keywords like `global`, `local`, and `const`. I—who usually write short codes for algorithm implementations and optimization modeling—think it is best to avoid same variables names in two different code blocks.

Some would write the above example as follows:

```
function f3(x)
  _a = 0
  return x + _a
```

3.7. Scope of Variables

```
end

a = 5
println(f3(1))
println(a)
```

where the underscore in front of a indicates it is a local variable.

Some others would write the above example as follows:

```
function f4(x, a)
    return x + a
end

a = 5
println(f4(1, a))
println(a)
```

which makes it clear that a is a function argument passed from outside the function block.

Suppose we are writing a code to compute the sum of a vector's elements. One may write:

```
a = [1 2 3 4 5]
s = 0
for i in 1:length(a)
    s += a[i]
end
```

If you run the above code in REPL, it will generate an error: "s not defined." In Julia, a `for` loop creates a scope block and there is an issue regarding global versus local variables. But the following code works fine:

```
function my_sum(a)
    s = 0
    for i in 1:5
        s += a[i]
    end
```

```
    return s
end

a = [1; 2; 3; 4; 5]
my_sum(a)
```

Since `function` has created a scope block already, the `for` loop inside is already part of the scope block.

For more detailed explanation and more examples, please consult with the official document[7].

3.8 Random Number Generation

In scientific computation, we often need to generate random numbers. Monte Carlo Simulation is such a case. We can generate a random number from the uniform distribution between 0 and 1 by simply calling `rand()`:

```
julia> rand()
0.8689474478700132

julia> rand()
0.33488929348173135
```

The generated number will be different for each call.

We can also create a vector of random numbers, for example of size 5:

```
julia> rand(5)
5-element Array{Float64,1}:
 0.848729
 0.18833
 0.591469
 0.59092
 0.0262999
```

or a matrix of random numbers, for example of size 4 by 3:

---

[7] https://docs.julialang.org/en/v1/manual/variables-and-scoping/

3.8. Random Number Generation

```
julia> rand(4,3)
4x3 Array{Float64,2}:
 0.34406    0.335058   0.261013
 0.34656    0.488157   0.600716
 0.0110059  0.0919956  0.501252
 0.894159   0.301035   0.4308
```

Random numbers from Uniform[0, 100]:

```
rand() * 100
```

A vector of n random numbers from Uniform$[a, b]$:

```
rand(n) * (b-a) + a
```

We can also use `rand()` for choosing an index randomly from a range:

```
julia> rand(1:10)
8

julia> rand(1:10)
2
```

Similarly, Julia provides a function called `randn()` for the standard Normal distribution with mean 0 and standard deviation 1, or $N(0, 1)$.

```
julia> randn(2,3)
2x3 Array{Float64,2}:
  2.31415   -0.309773  -0.0174724
 -0.316515   0.514558  -1.53451
```

To generate 10 random numbers from a general Normal distribution $N(\mu, \sigma^2)$:

```
julia> randn(10) .* sigma .+ mu
10-element Array{Float64,1}:
 52.2694
 51.4202
 45.4651
 51.9061
 51.1799
 47.272
 47.585
 48.4993
 54.2316
 46.5286
```

where `mu=50` and `sigma=3` are used. Since `randn(10)` is a vector and `sigma` and `mu` are scalars, vectorized operations `.*` and `.+` are used.

Perhaps, we can write our own function for $N(\mu, \sigma^2)$:

```
function my_randn(n, mu, sigma)
  return randn(n) .* sigma .+ mu
end
```

and call it:

```
julia> my_randn(10, 50, 3)
10-element Array{Float64,1}:
 48.0865
 47.7267
 47.3458
 48.2318
 53.5715
 54.3249
 53.3419
 46.0114
 49.7636
 56.0803
```

For any other advanced usages related to probabilistic distributions, the `StatsFuns`

3.8. Random Number Generation

package is available from the Julia Statistics group[8]. We need to first install the package:

```
julia> using Pkg
julia> Pkg.add("StatsFuns")
```

To use functions available in the `StatsFuns` package, we first load the package:

```
julia> using StatsFuns
```

For a Normal distribution with $\mu = 50$ and $\sigma = 3$, we set:

```
julia> mu = 50; sigma = 3;
```

The probability density function (PDF) value evaluated at 52:

```
julia> normpdf(mu, sigma, 52)
0.10648266850745075
```

The cumulative distribution function (CDF) value evaluated at 50:

```
julia> normcdf(mu, sigma, 50)
0.5
```

The inverse of CDF for probability 0.5:

```
julia> norminvcdf(mu, sigma, 0.5)
50.0
```

For many other probability distributions such as Binomial, Gamma, and Poisson distributions, similar functions are available from the `StatsFuns` package[9].

---

[8] http://juliastats.github.io
[9] https://github.com/JuliaStats/StatsFuns.jl

3.9 File Input/Output

A typical flow of computing in operations research, or any other computational sciences, is to read data from files, to run some computations, and to write the results on an output files for records. File read/write or input/output (I/O) is a very useful and important process. Suppose we have a data file named `data.txt` that looks like:

Listing 3.1: *code/chap3/data.txt*

```
This is the first line.
This is the second line.
This is the third line.
```

We can open the file and read all lines from the file as follows:

```julia
datafilename = "data.txt"
datafile = open(datafilename)
data = readlines(datafile)
close(datafile)
```

Then `data` is an array with each line being an element. Since `data` contains what we need, we close the file.

```julia
julia> data
3-element Array{String,1}:
 "This is the first line."
 "This is the second line."
 "This is the third line."
```

We usually access each line in a for-loop:

```julia
for line in data
  println(line)
  # process each line here...
end
```

3.9. File Input/Output

Writing the results to files is also similarly done.

```
outputfilename = "results1.txt"
outputfile = open(outputfilename, "w")
print(outputfile, "Magic Johnson")
println(outputfile, " is a basketball player.")
println(outputfile, "Michael Jordan is also a basketball player.")
close(outputfile)
```

The result looks like:

Listing 3.2: *code/chap3/results1.txt*

```
Magic Johnson is a basketball player.
Michael Jordan is also a basketball player.
```

By using "a" option, we can append to the existing file:

```
outputfilename = "results2.txt"
outputfile = open(outputfilename, "a")
println(outputfile, "Yuna Kim is a figure skating player.")
close(outputfile)
```

The result will read:

Listing 3.3: *code/chap3/results2.txt*

```
Magic Johnson is a basketball player.
Michael Jordan is also a basketball player.
Yuna Kim is a figure skating player.
```

The most important data format may be comma-separated-values files, called CSV files. Data in each line is separated by commas. For example, consider the simple network presented in Figure 3.1. This network may be represented as the following tabular data:

Chapter 3. Basics of the Julia Language

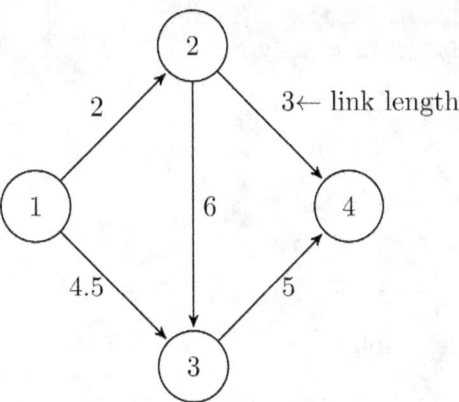

Figure 3.1: A simple network

start node	end node	link length
1	2	2
1	3	4.5
2	3	6
2	4	3
3	4	5

In spreadsheet software like Excel, you can enter as follows

	A	B	C
1	start node	end node	link length
2	1	2	2
3	1	3	4.5
4	2	3	6
5	2	4	3
6	3	4	5

and save it as 'data.csv'. When you save it in Excel, you need to choose 'Windows Comma Separated (.csv)', instead of just 'Comma Separated Values (.csv)'. CSV files are basically text files. The saved data.csv file will look like:

3.9. File Input/Output

Listing 3.4: *code/chap3/data.csv*

```
start node,end node,link length
1,2,2
1,3,4.5
2,3,6
2,4,3
3,4,5
```

We read this CSV file as follows:

```
using DelimitedFiles
csvfilename = "data.csv"
csvdata = readdlm(csvfilename, ',', header=true)
data = csvdata[1]
header = csvdata[2]
```

With the `header=true` option, we can separate the header from the data. We obtained `data` as an array of `Float64`:

```
julia> data
5x3 Array{Float64,2}:
 1.0  2.0  2.0
 1.0  3.0  4.5
 2.0  3.0  6.0
 2.0  4.0  3.0
 3.0  4.0  5.0
```

Since the third column represents link length, the type of `Float64` is fine. On the other hand, the first and second columns represent the node indices, which must be integers. We change the type:

```
julia> start_node = round.(Int, data[:,1])
5-element Array{Int64,1}:
 1
 1
 2
```

```
2
3

julia> end_node = round.(Int, data[:,2])
5-element Array{Int64,1}:
 2
 3
 3
 4
 4

julia> link_length = data[:,3]
5-element Array{Float64,1}:
 2.0
 4.5
 6.0
 3.0
 5.0
```

After some computation, suppose we obtained the following result:

```
value1 = [1.4; 3.1; 5.3; 2.7]
value2 = [4.3; 7.0; 3.6; 6.2]
```

which we want to save in the following format:

node	first value	second value
1	1.4	4.3
2	3.1	7.0
3	5.3	3.6
4	2.7	6.2

We can simply write the values on a file in the format we want:

```
resultfile = open("result.csv", "w")
println(resultfile, "node, first value, second value")
for i in 1:length(value1)
  println(resultfile, "$i, $(value1[i]), $(value2[i])")
end
close(resultfile)
```

3.10. Plotting

The result file looks like:

Listing 3.5: *code/chap3/result.csv*
```
node, first value, second value
1, 1.4, 4.3
2, 3.1, 7.0
3, 5.3, 3.6
4, 2.7, 6.2
```

3.10 Plotting

For plotting, I recommend the `PyPlot` package for most people. The `PyPlot` package calls the famous Python plotting module called `matplotlib.pyplot`. The power of PyPlot comes in at the cost of installing Python, which happens automatically. Read this Wiki page[10] for an introduction. You can use PyPlot via the `Plots` package.

At this moment, I find the `PyPlot` package provides more powerful plotting tools that are suitable for generating plots for academic research papers—yes, Python has been around for a while. It just comes at the cost of installing additional software packages.

3.10.1 The `PyPlot` Package

To use `PyPlot`, we need a distribution of Python and matplotlib. Julia can automatically install a distribution for you, which will be private to Julia and will not be accessible outside of Julia. If you want to use your existing Python, please consult with the documentation of `PyCall`[11].

The following process will automatically install Python and matplotlib.

1. Open a new terminal window and run Julia. Initialize the `PYTHON` environment variable:

---

[10] https://en.wikibooks.org/wiki/Introducing_Julia/Plotting
[11] https://github.com/JuliaPy/PyCall.jl

```
julia> ENV["PYTHON"] = ""
""
```

2. Install PyPlot:

```
julia> using Pkg
julia> Pkg.add("PyPlot")
```

An example is given below:

Listing 3.6: *code/chap3/plot1.jl*

```julia
using PyPlot

# Preparing a figure object
fig = figure()

# Data
x = range(0, stop=2*pi, length=1000)
y = sin.(3*x)

# Plotting with linewidth and linestyle specified
plot(x, y, color="blue", linewidth=2.0, linestyle="--")

# Labeling the axes
xlabel(L"value of $x$")
ylabel(L"\sin(3x)")

# Title
title("Test plotting")

# Save the figure as PNG and PDF
savefig("plot1.png")
savefig("plot1.pdf")

# Close the figure object
close(fig)
```

The result is:

3.10. Plotting

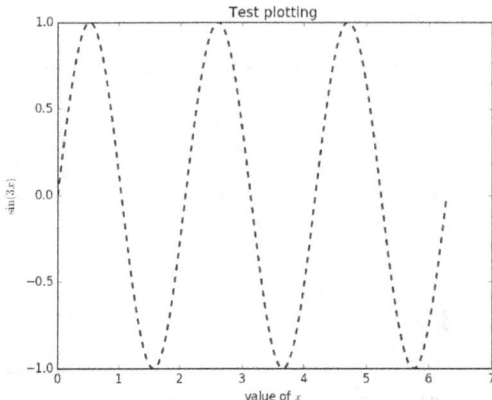

When you use PyPlot for the first time, it will take some time to precompile. In the above code, L in front of `"value of $x$"` means that the string inside the quotation marks will be LaTeX strings. When you have dollar signs ($) inside L"...", it is a combination of text and math symbols; when you don't, like as in L"\sin(3x)", the entire text consists of math symbols only.

For another example, suppose that you obtained lower bounds and upper bounds from iterations of an algorithm. Then you want to put these two data in a single plot.

Listing 3.7: *code/chap3/plot2.jl*

```julia
using PyPlot

# Data
lower_bound = [4.0, 4.2, 4.4, 4.8, 4.9, 4.95, 4.99, 5.00]
upper_bound = [5.4, 5.3, 5.3, 5.2, 5.2, 5.15, 5.10, 5.05]
iter = 1:8

# Creating a new figure object
fig = figure()

# Plotting two datasets
plot(iter, lower_bound, color="red", linewidth=2.0, linestyle="-",
 marker="o", label=L"Lower Bound $Z^k_L$")
plot(iter, upper_bound, color="blue", linewidth=2.0, linestyle="-.",
```

```
 marker="D", label=L"Upper Bound $Z^k_U$")

# Labeling axes
xlabel(L"iteration clock $k$", fontsize="xx-large")
ylabel("objective function value", fontsize="xx-large")

# Putting the legend and determining the location
legend(loc="upper right", fontsize="x-large")

# Add grid lines
grid(color="#DDDDDD", linestyle="-", linewidth=1.0)
tick_params(axis="both", which="major", labelsize="x-large")

# Title
title("Lower and Upper Bounds")

# Save the figure as PNG and PDF
savefig("plot2.png")
savefig("plot2.pdf")

# Closing the figure object
close(fig)
```

The result is:.

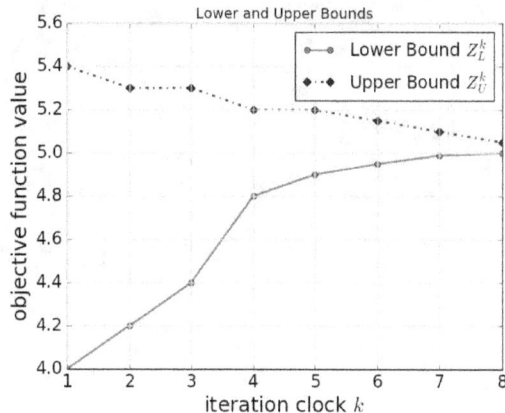

One can also create a histogram easily.

3.10. Plotting

Listing 3.8: *code/chap3/plot3.jl*

```julia
using PyPlot

# Data
data = randn(100)  # Some Random Data
nbins = 10         # Number of bins

# Creating a new figure object
fig = figure()

# Histogram
plt[:hist](data, nbins)

# Title
title("Histogram")

# Save the figure as PNG and PDF
savefig("plot3.png")
savefig("plot3.pdf")

# Closing the figure object
close(fig)
```

The result is:

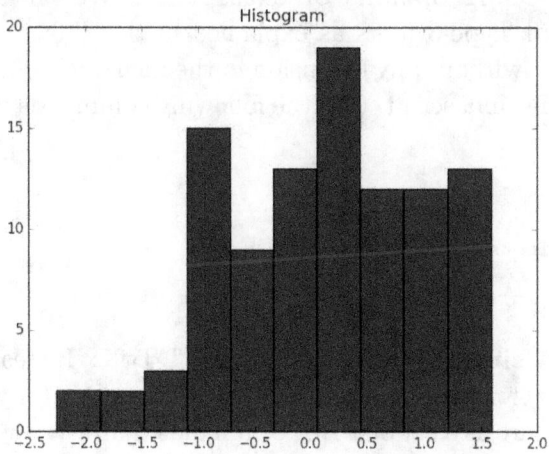

Many examples and sample codes for using `PyPlot` are provided in the following link:

- Various Julia plotting examples using PyPlot[12]

Since `PyPlot` calls functions from `matplotlib`, the following documents are also helpful.

- Matplotlib Beginner's Guide[13]

- Matplotlib Examples[14]

3.10.2 Avoiding Type-3 Fonts in `PyPlot`

Some journal submission systems don't like PDF files saved by matplotlib. It is usually because the system cannot handle some newer font styles, Type-3 fonts. There are two ways:

---

[12]https://gist.github.com/gizmaa/7214002
[13]http://matplotlib.org/users/beginner.html
[14]http://matplotlib.org/examples/index.html

3.10. Plotting

- **Method 1. Using options of `matplotlib`:** We can specifically tell matplotlib to avoid Type-3 fonts as explained in this link.[15] Create a file called `matplotlibrc` (without any extension in the filename) and place it in the same directory as the Julia script. Put the following commands in the `matplotlibrc` file:

```
ps.useafm          : True
pdf.use14corefonts : True
text.usetex        : True
```

- **Method 2. Using the `pgf` package of LaTeX:** Instead of using the command `savefig("myplot.pdf")`, one can use `savefig("myplot.pgf")`, which will save the figure as a set of LaTeX commands that uses the `pgf` package. Then use the following command to include the figure in the main LaTeX document:

```
\begin{figure} \centering
\resizebox{0.7\textwidth}{!}{\input{myplot.pgf}}
\caption{Figure caption goes here..}
\label{fig:myplot}
\end{figure}
```

Don't forget to include \usepackage{pgf} in the preamble of your main LaTeX document.

---

[15] http://nerdjusttyped.blogspot.com/2010/07/type-1-fonts-and-matplotlib-figures.html

4

Selected Topics in Numerical Methods

Although this book does not aim to cover details of numerical methods and algorithms, this chapter will go over very basics of selected topics. Namely, curve fitting, numerical differentiation, and numerical integration are briefly explained. While students and researchers in operations research may not directly use these methods in their own problem solving, the concepts behind these fundamental topics are often useful to understand more complicated and advanced methods. It is also helpful to recognize how related computer software for numerical computations would be designed and what the limitations are.

4.1 Curve Fitting

From data collection or experiment results, we may obtain discrete data sets such as

$$(x_1, y_1), (x_2, y_2), ..., (x_n, y_n)$$

where x_i is an input value and y_i is an output value for each $i = 1, ..., n$. For example, x_i could be the price of a popular book in time period i, and y_i is the corresponding sales volume in time period i. Instead of having this discrete data set, we often want to represent the relationship between x and y as an analytical expression, such as a linear function:

$$y = \beta_1 + \beta_2 x$$

4.1. Curve Fitting

or an exponential function
$$y = \beta_1 e^{\beta_2 x}$$
In general, a mathematical function with a vector of parameters β
$$y = f(x; \beta)$$
that makes sense in the context of x, y, and the relationship. *Curve fitting* aims to find the values of parameters used in the function form $f(\cdot; \beta)$ so that the obtained analytical functional form is the closest to the original discrete data set.

Related to curve fitting, *interpolation* finds the values of parameters so that the analytical function *passes through* the discrete data points. This approach assumes that the discrete data set is accurate and exact. When we use polynomial functions for interpolation—called 'polynomial interpolation'—available methods are Lagrange's Method, Newton's Method, and Neville's Method.

Instead of using only one polynomial function for the entire data set, we can use a piecewise polynomial function, a function whose segments are separate polynomial functions. The most popular choice is to use a piecewise cubic function, or cubic spline, and the method is called 'cubic spline interpolation'. For interpolation methods, most books with 'numerical methods' or 'numerical analysis' in the title are helpful; see Kiusalaas (2013)[1] for example.

In curve fitting, the objective is to determine β in $f(\cdot; \beta)$ to match the values of $f(x_i; \beta)$ to y_i for all $i = 1, ..., n$ as much as we can. The definition of the best match or best fit depends on one's definition. The most popular definition is based on the least-squares. That is, we aim to minimize

$$S(\beta) = \sum_{i=1}^{n} (y_i - f(x_i; \beta))^2$$

by optimally choosing β. This is a nonlinear optimization problem in general. When we want a linear function for $f(x)$—called linear regression—the problem becomes a quadratic optimization problem, which can be solved relatively easily.

Finding an optimal β is related to solving a system of equations:

$$\frac{\partial S}{\partial \beta_i} = 0 \quad \forall i = 1, ..., m$$

---

[1] Kiusalaas, J., 2013. Numerical methods in engineering with Python 3. Cambridge university press.

where m is the number of parameters. In case of linear regression, the problem is to solve a system of linear equations.

For general nonlinear least-squares fit, the Levenberg-Marquardt algorithm is popular. See Nocedal and Wright (2006)[2] for details. In Julia, the LsqFit package[3] from the JuliaOpt group implements the Levenberg-Marquardt algorithm. First add the package:

```
julia> using Pkg
julia> Pkg.add("LsqFit")
```

Suppose we have the following data set:

```
xdata = [ 15.2;  19.9;   2.2;  11.8;  12.1;  18.1;  11.8;  13.4;  11.5;   0.5;
          18.0;  10.2;  10.6;  13.8;   4.6;   3.8;  15.1;  15.1;  11.7;   4.2 ]
ydata = [  0.73;  0.19;  1.54;  2.08;  0.84;  0.42;  1.77;  0.86;  1.95;  0.27;
           0.39;  1.39;  1.25;  0.76;  1.99;  1.53;  0.86;  0.52;  1.54;  1.05 ]
```

We would like to use the following function form:

$$f(x) = \beta_1 \left(\frac{x}{\beta_2}\right)^{\beta_3-1} \exp\left(-\left(\frac{x}{\beta_2}\right)^{\beta_3}\right)$$

To determine β, we prepare a curve fitting model as a function that returns a vector:

```
function model(xdata, beta)
  values = similar(xdata)
  for i in 1:length(values)
    values[i] = beta[1] * ((xdata[i]/beta[2])^(beta[3]-1)) *
                (exp( - (xdata[i]/beta[2])^beta[3] ))
  end
  return values
end
```

This can be equivalently written as:

---

[2]Nocedal, J. and Wright, S., 2006. Numerical optimization. Springer Science & Business Media.
[3]https://github.com/JuliaNLSolvers/LsqFit.jl

4.1. Curve Fitting

```
model(x,beta) = beta[1] * ((x/beta[2]).^(beta[3]-1)) .*
                          (exp.( - (x/beta[2]).^beta[3] ))
```

where `.^` and `.*` represent element-wise operations.

With some initial guess on β as [3.0, 8.0, 3.0], we do

```
using LsqFit
fit = curve_fit(model, xdata, ydata, [3.0, 8.0, 3.0])
```

The obtained parameter values are accessed by:

```
julia> beta = fit.param
3-element Array{Float64,1}:
  4.459414325729536
 10.254403821607001
  1.8911376587646551
```

and the error estimates for fitting parameters are accessed by:

```
julia> margin_error(fit)
3-element Array{Float64,1}:
 1.240597479103558
 1.410107749179365
 0.4015751605563139
```

The result of curve fitting is presented in Figure 4.1. The complete code is provided:

Listing 4.1: Curve Fitting
code/chap4/curve_fit.jl

```
using LsqFit    # for curve fitting
using PyPlot    # for drawing plots

# preparing data for fitting
xdata = [ 15.2; 19.9;  2.2; 11.8; 12.1; 18.1; 11.8; 13.4; 11.5;  0.5;
```

Chapter 4. Selected Topics in Numerical Methods

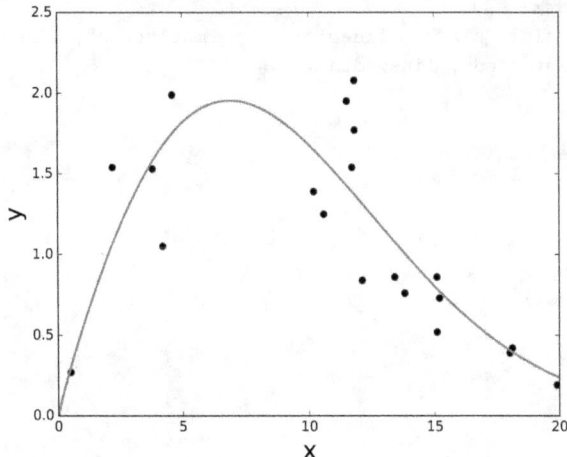

Figure 4.1: Curve Fitting Result

```
            18.0; 10.2; 10.6; 13.8;  4.6;  3.8; 15.1; 15.1; 11.7;  4.2 ]
ydata = [ 0.73; 0.19; 1.54; 2.08; 0.84; 0.42; 1.77; 0.86; 1.95; 0.27;
          0.39; 1.39; 1.25; 0.76; 1.99; 1.53; 0.86; 0.52; 1.54; 1.05 ]

# defining a model
model(x,beta) = beta[1] * ((x/beta[2]).^(beta[3]-1)) .*
                        (exp.( - (x/beta[2]).^beta[3] ))

# run the curve fitting algorithm
fit = curve_fit(model, xdata, ydata, [3.0, 8.0, 3.0])

# results of the fitting
beta_fit = fit.param
errors = margin_error(fit)

# preparing the fitting evaluation
xfit = 0:0.1:20
yfit = model(xfit, fit.param)

# Creating a new figure object
fig = figure()
```

4.2. Numerical Differentiation

```
# Plotting two datasets
plot(xdata, ydata, color="black", linewidth=2.0, marker="o", linestyle="None")
plot(xfit, yfit, color="red", linewidth=2.0)

# Labeling axes
xlabel("x", fontsize="xx-large")
ylabel("y", fontsize="xx-large")

# Save the figure as PNG and PDF
savefig("fit_plot.png")
savefig("fit_plot.pdf")

# Closing the figure object
close(fig)
```

4.2 Numerical Differentiation

Given a function $f(x)$, we often need to compute its derivative without actually differentiating the function. That is, without the analytical form expressions of $f'(x)$ or $f''(x)$, we need to compute them numerically. This process of *numerical differentiation* is usually done by *finite difference approximations*.

The idea is simple. The definition of the first-order derivative is

$$f'(x) = \lim_{h \to 0} \frac{f(x+h) - f(x)}{h}$$

from which we obtain a finite difference approximation:

$$f'(x) \approx \frac{f(x+h) - f(x)}{h}$$

for sufficiently small $h > 0$. This approximation is called the *forward* finite difference approximation. The *backward* approximation is

$$f'(x) \approx \frac{f(x) - f(x-h)}{h}$$

and the *central* approximation is

$$f'(x) \approx \frac{f(x+h) - f(x-h)}{2h}.$$

Suppose we have discrete points $x_1, x_2, ..., x_n$. For mid-points from x_2 to x_{n-1}, we can use any finite difference approximation and typically prefer the central approximation. At the boundary points x_1 and x_n, the central approximation is unavailable; hence we need to use the forward approximation for x_1 and the backward approximation for x_n.

To be more precise, let us consider the following Taylor series expansions:

$$f(x+h) = f(x) + hf'(x) + \frac{h^2}{2!}f''(x) + \frac{h^3}{3!}f'''(x) + \frac{h^4}{4!}f^{(4)}(x) + \cdots \quad (4.1)$$

$$f(x-h) = f(x) - hf'(x) + \frac{h^2}{2!}f''(x) - \frac{h^3}{3!}f'''(x) + \frac{h^4}{4!}f^{(4)}(x) - \cdots \quad (4.2)$$

From (4.1), we obtain the forward approximation:

$$f'(x) = \frac{f(x+h) - f(x)}{h} + \frac{h}{2!}f''(x) + \frac{h^2}{3!}f'''(x) + \frac{h^3}{4!}f^{(4)}(x) + \cdots$$
$$= \frac{f(x+h) - f(x)}{h} + \mathcal{O}(h)$$

From (4.2), we obtain the backward approximation:

$$f'(x) = \frac{f(x) - f(x-h)}{h} + \mathcal{O}(h)$$

By subtracting (4.2) from (4.1), we obtain the central approximation

$$f'(x) = \frac{f(x+h) - f(x-h)}{2h} + \mathcal{O}(h^2)$$

Note that the central approximation has a higher-order truncation error $\mathcal{O}(h^2)$ than the forward and backward approximations, which means typically smaller errors for sufficiently small h.

The second-order derivative is similarly approximated. By adding (4.1) and (4.2), we obtain

$$f(x+h) + f(x-h) = 2f(x) + h^2 f''(x) + \frac{h^4}{12}f^{(4)}(x) + \cdots$$

4.2. Numerical Differentiation

which leads to the central finite difference approximation of the second-order derivative:

$$f''(x) = \frac{f(x+h) - 2f(x) + f(x-h)}{h^2} + \frac{h^2}{12}f^{(4)}(x) + \cdots$$
$$= \frac{f(x+h) - 2f(x) + f(x-h)}{h^2} + \mathcal{O}(h^2)$$

We can similarly derive the forward and backward approximations of the second-order derivative. We can also approximate the third- and fourth-order derivatives. See Kiusalaas (2013)[4] for details.

In Julia, the `Calculus` package is available for numerical differentiation. First install and import the package:

```
using Pkg
Pkg.add("Calculus")
using Calculus
```

Suppose we want to differentiate the following function:

$$f(x) = x^3 e^x + \sin x$$

We prepare the function in Julia:

```
f(x) = x^3 * exp(x) + sin(x)
```

The first-order derivative of f at 1.0:

```
julia> derivative(f, 1.0)
11.413429620197812
```

and the second-order derivative of f at 1.0:

---

[4]Kiusalaas, J., 2013. Numerical methods in engineering with Python 3. Cambridge university press.

```
julia> second_derivative(f, 1.0)
34.49618758929225
```

For functions with multiple variables, we compute the gradient and hessian. We consider
$$g(x) = (x_1)^2 \sin(3x_2) + e^{-2x_3}$$
and prepare the function in Julia:

```
g(x) = (x[1])^2 * sin(3x[2]) + exp(-2x[3])
```

The gradient at $[3.0, 1.0, 2.0]$:

```
julia> Calculus.gradient(g, [3.0, 1.0, 2.0])
3-element Array{Float64,1}:
   0.8467200483621847
 -26.729797406378975
  -0.03663127778181647
```

and the Hessian at the same point:

```
julia> hessian(g, [3.0, 1.0, 2.0])
3x3 Array{Float64,2}:
   0.282241  -17.8199    0.0
 -17.8199    -11.4307    0.0
   0.0        0.0        0.0732632
```

The Calculus package[5] also offers symbolic differentiation.

4.3 Numerical Integration

Numerical integration of a continuous function is to approximate
$$I = \int_a^b f(x)\,\mathrm{d}x$$

---
[5] https://github.com/JuliaMath/Calculus.jl

4.3. Numerical Integration

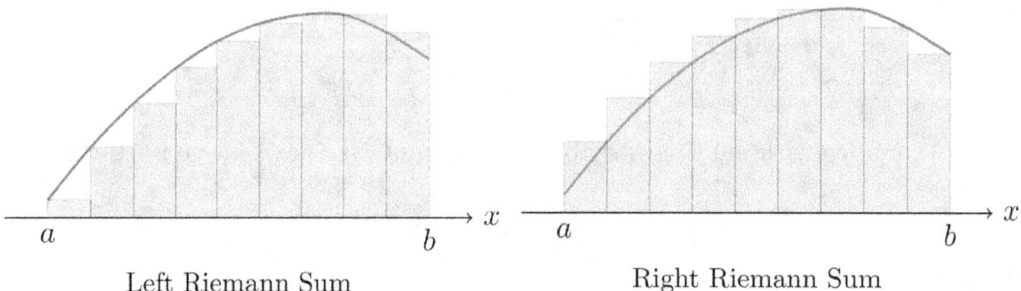

Figure 4.2: Riemann Sum

The most obvious way of approximating this Riemann integral is using the Riemann sum. With mesh points $x_1 < x_2 < ... < x_n$ with $x_1 = a$ and $x_n = b$, we may write the Riemann sum as follows:

$$R = \sum_{i=1}^{n-1} f(t_i)(x_{i+1} - x_i)$$

where $t_i \in [x_i, x_{i+1}]$ is an evaluation point in each sub-interval. If $t_i = x_i$, then R is a left Riemann sum, and if $t_i = x_{i+1}$, then R is a right Riemann sum. If $t_i = (x_i + x_{i+1})/2$, then R is a middle Riemann sum.

In case of the example shown in Figure 4.2, the left Riemann sum underestimates the integral, while the right Riemann sum overestimates (but not always). One may think the average of two would be a good approximation. It is called the trapezoidal sum. That is:

$$\frac{1}{2}\sum_{i=1}^{n-1} f(x_i)(x_{i+1} - x_i) + \frac{1}{2}\sum_{i=1}^{n-1} f(x_{i+1})(x_{i+1} - x_i) = \sum_{i=1}^{n-1} \frac{f(x_i) + f(x_{i+1})}{2}(x_{i+1} - x_i)$$

When $x_{i+1} - x_i = h$ for all i, then

$$\sum_{i=1}^{n-1} \frac{f(x_i) + f(x_{i+1})}{2} h = \sum_{i=1}^{n} w_i f(x_i) h$$

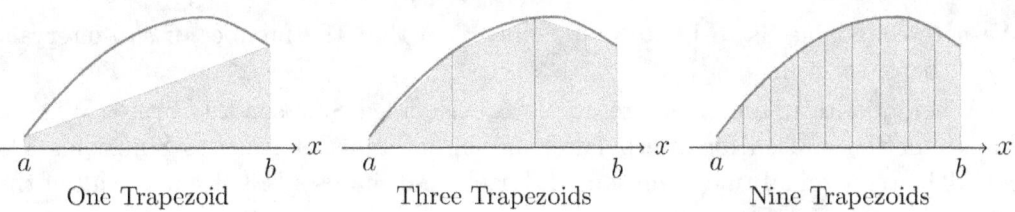

Figure 4.3: Approximation by Trapezoids

where
$$w_i = \begin{cases} 1/2 & \text{if } i = 1 \\ 1 & \text{if } i = 2, ..., n-1 \\ 1/2 & \text{if } i = n \end{cases}$$

which is the trapezoidal rule for numerical integration.

The trapezoidal rule can be viewed in another perspective. We approximate function $f(x)$ by a straight line in each sub-interval $[x_i, x_{i+1}]$; then the area in the interval looks like a trapezoid. See Figure 4.3

Simpson's rule approximates function $f(x)$ by a quadratic function instead of a straight line. Suppose we have evenly spaced mesh points $a = x_1 < x_2 < ... < x_n = b$. For an arbitrary interval $[x_i, x_{i+2}]$, we use three points—x_i, x_{i+2}, and the midpoint $x_{i+1} = (x_i + x_{i+2})/2$—to interpolate $f(x)$ by a quadratic function. We approximate

$$\int_{x_i}^{x_{i+2}} f(x)\,dx \approx \frac{1}{3}\Big[f(x_i) + 4f(x_{i+1}) + f(x_{i+2})\Big]h$$

For the entire interval we do:

$$\int_a^b f(x)\,dx = \sum_{i=1,3,5,...}^{n-2} \int_{x_i}^{x_{i+2}} f(x)$$

which leads to:

$$\int_a^b f(x)\,dx \approx \frac{1}{3}\Big[f(x_1) + 4f(x_2) + 2f(x_3) + 4f(x_4) + ... \\ + 2f(x_{n-2}) + 4f(x_{n-1}) + f(x_n)\Big]h \quad (4.3)$$

4.3. Numerical Integration

We note that n needs to be an odd number, so that the number of sub-intervals, $n-1$, becomes even.

When quadratic functions are used, (4.3) is called Simpson's 1/3 rule. When a cubic function is used for interpolation in sub-intervals, it leads to Simpson's 3/8 rule. The trapezoidal rule, Simpson's 1/3 rule, and Simpson's 3/8 rule are all of the form:

$$S = \sum_{i=1}^{n} w_i f(x_i) h$$

with different rules to define weights w_i.

The `quadgk()` function from the `QuadGK` package provides a method for numerical integration; in particular, Gauss-Kronrod integration method. Add the package as follows:

```
julia> using Pkg
julia> Pkg.add("QuadGK")
```

Suppose we integrate the following function over the interval $[0.0, 1.0]$:

$$f(x) = -\cos 3x + x^2 e^{-x}$$

which is in Julia:

```
f(x) = - cos(3x) + x^2 * exp(-x)
```

The integration can be done as follows:

```
julia> using QuadGK
julia> quadgk(f, 0.0, 1.0)
(0.11356279145616598, 2.123301534595612e-14)
```

where the first output is the integral value and the second output is the error.

4.4 Automatic Differentiation

While numerical differentiation based on finite differences has been popularly used, there is an issue of computational errors from finite differencing. With the developments of modern computer languages, a new paradigm on differentiation using computers has risen: automatic differentiation (AD). We should note that AD computes the derivatives exactly at no significant computational efforts—amazing! Also note that AD is not symbolic differentiation. Some computer programs like Mathematica and Maple, even MATLAB and Julia, can do symbolic differentiation, in the same way that humans do differentiation.

To understand how AD works, let us consider ordered pairs in the form of:

$$\vec{x} = (x, x'), \qquad \vec{y} = (y, y')$$

We define operations between these two pairs:

$$\vec{x} + \vec{y} = (x, x') + (y, y') = (x + y, x' + y')$$
$$\vec{x} - \vec{y} = (x, x') - (y, y') = (x - y, x' - y')$$
$$\vec{x} \times \vec{y} = (x, x') \times (y, y') = (xy, xy' + x'y)$$
$$\frac{\vec{x}}{\vec{y}} = \frac{(x, x')}{(y, y')} = \left(\frac{x}{y}, \frac{x'y - xy'}{y^2}\right)$$

In the third operation for $\times$, look at the second term in the result: $xy' + x'y$, which is exactly the derivative of xy. In the fourth operation, the second term represents the derivative of x/y. The above operations are defined so that the second term represents the derivative of the first term in the result.

Let's consider an example now. Given a function:

$$f(x) = \frac{(x-1)(x+3)}{x}$$

and we want to compute $f(2)$ and $f'(2)$. (Note, we don't compute $f(x)$ and $f'(x)$; we compute for a specific $x = 2$.) We do the following operations:

$$\vec{f}(x) = \frac{(\vec{x} - \vec{1})(\vec{x} + \vec{3})}{\vec{x}}$$
$$= \frac{((x, 1) - (1, 0)) \times ((x, 1) + (3, 0))}{(x, 1)}$$

4.4. Automatic Differentiation

where we use $\vec{x} = (x, 1)$ since $dx/dx = 1$ and $\vec{c} = (c, 0)$ for any constant c. When $x = 2$, this leads to:

$$\vec{f}(2) = \frac{((2,1) - (1,0)) \times ((2,1) + (3,0))}{(2,1)}$$

$$= \frac{(1,1) \times (5,1)}{(2,1)}$$

$$= \frac{(5,6)}{(2,1)}$$

$$= \left(\frac{5}{2}, \frac{7}{4}\right)$$

Direct computations tell us that $f(2) = 5/2$ and $f'(2) = 7/4$, which match the results obtained by AD with the ordered pairs.

Using the chain rule, we can define more operations such as:

$$\sin \vec{x} = \sin(x, x') = (\sin x, x' \cos x)$$
$$\cos \vec{x} = \cos(x, x') = (\cos x, -x' \sin x)$$
$$\exp \vec{x} = \exp(x, x') = (\exp x, x' \exp x)$$
$$\log \vec{x} = \log(x, x') = (\log x, x'/x)$$

and the list goes on.

To use AD, the computer needs to know how the function is defined. For example, in Julia the above function $f(x)$ is writte as:

```
function f(x)
    return (x-1) * (x+3) / x
end
```

Modern computer languages like Julia can recognize each operator like -, +, *, and / and substitute each operator with the operation for ordered pairs as defined in AD.

There is the JuliaDiff group[6] who implements simple to advanced techniques of AD. The codes from the JuliaDiff group is used in the Optim and JuMP packages.

Just to see how we can use AD in Julia, first install the ForwardDiff package:

---
[6] http://www.juliadiff.org

```
using Pkg
Pkg.add("ForwardDiff")
```

With the function definition of f, we can compute $f'(2)$ as follows:

```
using ForwardDiff
g = x -> ForwardDiff.derivative(f, x)
g(2)
```

For a vector function:
$$f(x) = (x_1 - 2)e^{x_2} - \sin x_3$$
we can do as follows:

```
using ForwardDiff

f(x) = (x[1]-2) * exp(x[2]) - sin(x[3])

g = x -> ForwardDiff.gradient(f, x)
h = x -> ForwardDiff.hessian(f, x)

g([3.0, 2.0, 1.0])
h([3.0, 2.0, 1.0])
```

The result is:

```
julia> g([3.0, 2.0, 1.0])
3-element Array{Float64,1}:
  7.38905609893065
  7.38905609893065
 -0.5403023058681398

julia> h([3.0, 2.0, 1.0])
3×3 Array{Float64,2}:
 0.0      7.38906  0.0
 7.38906  7.38906  0.0
 0.0      0.0      0.841471
```

4.4. Automatic Differentiation

5 The Simplex Method

This chapter aims to implement the Simplex Method to solve Linear Programming (LP) problems. Of course, solvers for LP problems are already available through many Julia packages such as `Clp`, `CPLEX`, and `Gurobi`. The main purpose of creating our own codes for the Simplex Method is to learn some advanced features of Julia and practice Julia programming. Readers can skip this chapter and proceed to the next chapters.

This chapter assumes prior knowledge of the Simplex Method and familiarity with the Simplex Method in tableau form. A course in (Deterministic) Operations Research or Linear Programming, at either undergraduate or graduate level, would be sufficient.

5.1 A Brief Description of the Simplex Method

This section provides a brief description of the Simplex Method. As different authors use different notations, we need to introduce the notation that this chapter use. We will use the notation of Bazaraa et al. (2011)[1] with slight changes, of which this section provides a summary. We consider an LP problem of the following form:

---

[1] Bazaraa, M.S., Jarvis, J.J. and Sherali, H.D., 2011. Linear Programming and Network Flows. John Wiley & Sons.

5.1. A Brief Description of the Simplex Method

$$\begin{aligned} \min \quad & \mathbf{c}^\top \mathbf{x} \\ \text{s.t.} \quad & \mathbf{A}\mathbf{x} = \mathbf{b} \\ & \mathbf{x} \geq \mathbf{0} \end{aligned}$$

where $\mathbf{A}$ is an $m \times n$ matrix with rank m and all other vectors have appropriate dimensions. The only different between our notation and the notation of Bazaraa et al. (2011) is that $\mathbf{c}$ is a column vector instead of a row vector.

Suppose we have a basis $\mathbf{B}$ and $\mathbf{A}$ can be decomposed as $\mathbf{A} = [\mathbf{B}\ \mathbf{N}]$. Accordingly, any $\mathbf{x}$ can be written as

$$\mathbf{x} = \begin{bmatrix} \mathbf{x}_B \\ \mathbf{x}_N \end{bmatrix} = \begin{bmatrix} \mathbf{B}^{-1}\mathbf{b} \\ \mathbf{0} \end{bmatrix}$$

The cost vector $\mathbf{c}$ can also be written as

$$\mathbf{c} = \begin{bmatrix} \mathbf{c}_B \\ \mathbf{c}_N \end{bmatrix}$$

so that the objective function value at the current solution is

$$z_0 = \mathbf{c}_B^\top \mathbf{x}_B.$$

If $\mathbf{x}_B = \mathbf{B}^{-1}\mathbf{b} \geq \mathbf{0}$, the current solution $\mathbf{x}$ is called a basic feasible solution (BFS). It is well-known that a BFS is equivalent to an extreme point of the feasible region. Furthermore, we know that there exists an optimal solution to the LP problem at an extreme point of the feasible region; therefore we can search the extreme points wisely, which is the essence of the Simplex Method.

We consider the linear system $\mathbf{A}\mathbf{x} = \mathbf{b}$, which can be written as

$$\mathbf{A}\mathbf{x} = [\mathbf{B}\ \mathbf{N}] \begin{bmatrix} \mathbf{x}_B \\ \mathbf{x}_N \end{bmatrix} = \mathbf{b}$$

Let J denote the current set of indices for the non-basic variables $\mathbf{x}_N$, $\mathbf{a}_j$ denote the j-th column vector of matrix $\mathbf{A}$, and $\mathbf{y}_j = \mathbf{B}^{-1}\mathbf{a}_j$. Then we can write

$$\begin{aligned} \mathbf{x}_B &= \mathbf{B}^{-1}\mathbf{b} - \mathbf{B}^{-1}\mathbf{N}\mathbf{x}_N \\ &= \mathbf{B}^{-1}\mathbf{b} - \sum_{j \in J} \mathbf{B}^{-1}\mathbf{a}_j x_j \end{aligned}$$

$$= \bar{\mathbf{b}} - \sum_{j \in J} \mathbf{y}_j x_j$$

where we defined $\bar{\mathbf{b}} = \mathbf{B}^{-1}\mathbf{b}$. This equation provides a way to write the original LP problem in terms of the non-basic variable $\mathbf{x}_N$. Note that the non-basic variable x_j for each $j \in J$ is currently set to zero. If we increase x_j, then $\mathbf{x}_B$ will change. Using column vectors $\mathbf{y}_j$ for all $j = 1, ..., m$, we can construct a matrix $\mathbf{Y}$.

The objective function can also be written in terms of $\mathbf{x}_N$ as follows:

$$\begin{aligned}
z &= \mathbf{c}^\top \mathbf{x} \\
&= \mathbf{c}_B^\top \mathbf{x}_B + \mathbf{c}_N^\top \mathbf{x}_N \\
&= \mathbf{c}_B^\top \left(\mathbf{B}^{-1}\mathbf{b} - \sum_{j \in J} \mathbf{B}^{-1}\mathbf{a}_j x_j \right) + \sum_{j \in J} c_j x_j \\
&= \mathbf{c}_B^\top \mathbf{B}^{-1}\mathbf{b} - \sum_{j \in J} (\mathbf{c}_B^\top \mathbf{B}^{-1}\mathbf{a}_j - c_j) x_j \\
&= z_0 - \sum_{j \in J} (z_j - c_j) x_j
\end{aligned}$$

where we defined $z_j = \mathbf{c}_B^\top \mathbf{B}^{-1}\mathbf{a}_j$ for each $j \in J$.

It is easy to observe that the current BFS is optimal, if $z_j - c_j \leq 0$ for all $j \in J$, since increasing any x_j cannot further decrease the objective function. On the other hand if $z_j - c_j > 0$ for some $j \in J$, we can increase the corresponding x_j to decrease the objective function.

The Simplex Method is based on pivoting, which is a process that changes the current basis to another with maintaining the feasibility. The pivoting process requires a variable that enters the basis (entering variable) and another variable that exits the basis (exiting variable). To determine an entering variable, we can choose among non-basic variables: pick $k \in J$ with $z_k - c_k > 0$.

To determine the exiting variable, we need to consider feasibility. With the entering variable is determined as x_k, we will be increasing x_k to a certain positive value. All other non-basic variables will remain at zero. Since our new $\mathbf{x}$ needs to be feasible, we need to enforce:

$$\mathbf{x}_B = \bar{\mathbf{b}} - \mathbf{y}_k x_k \geq \mathbf{0} \tag{5.1}$$

5.2. Searching All Basic Feasible Solutions

We will increase x_k up to the largest value that satisfies the above condition. Let i-th elements of $\mathbf{x}_B$, $\bar{\mathbf{b}}$, and $\mathbf{y}_k$ by x_{Bi}, $\bar{b}_i$, and y_{ik} respectively. The above condition can be written as:

$$\begin{bmatrix} x_{B1} \\ \vdots \\ x_{Br} \\ \vdots \\ x_{Bm} \end{bmatrix} = \begin{bmatrix} \bar{b}_1 \\ \vdots \\ \bar{b}_r \\ \vdots \\ \bar{b}_m \end{bmatrix} - \begin{bmatrix} y_{1k} \\ \vdots \\ y_{rk} \\ \vdots \\ y_{mk} \end{bmatrix} x_k \geq \begin{bmatrix} 0 \\ \vdots \\ 0 \\ \vdots \\ 0 \end{bmatrix}$$

First note that $\bar{b}_i \geq 0$ for all $i = 1, ..., m$. If $y_{ik} \leq 0$ for all $i = 1, ..., m$, we can make x_k arbitrarily large; therefore the problem is *unbounded*. If there is any row with $y_{ik} > 0$, we cannot increase x_k arbitrarily large. The value of x_k needs to be bounded by the ratio $\frac{\bar{b}_i}{y_{ik}}$. Therefore x_k is determined by the following *minimum ratio*:

$$x_k = \frac{\bar{b}_r}{y_{rk}} = \min_{i=1,...,m} \left\{ \frac{\bar{b}_i}{y_{ik}} : y_{ik} > 0 \right\}$$

This is called the min-ratio test. As a result, we see the r-th row is the critical row and determine the corresponding x_{Br} as the exiting variable.

Once we determined the entering variable x_k and the exiting variable x_{Br}, we form a new basic variable $\mathbf{x}_B$ and the new basis $\mathbf{B}$, and repeat the process until we find an optimal solution. In the subsequent sections, we will practice Julia programming by implementing the Simplex Method.

5.2 Searching All Basic Feasible Solutions

Before we develop a code for the Simplex Method, as an easier practice, let's try to search all BFS. Of course, this is an inefficient method, since the number of basic solutions can be very large. For example if the dimension of $\mathbf{A}$ is $m \times n$, then we can choose m from n to form a basis. For example, if $m = 3$ and $n = 10$, then $\binom{n}{m} = \binom{20}{5} = 15,504$. For any LP problem of practical size, searching all BFS is not a valid idea. But, we want some practice of Julia programming.

We will create a Julia function of the following form:

```
function search_BFS(c, A, b)
    ...
    return opt_x, obj
end
```

which accepts arrays c, A, and b as input arguments, and returns opt_x as the optimal x^* and obj as the optimal objective function value. For example, we have an instance of the following LP problem:

```
c = [-3; -2; -1; -5; 0; 0; 0]
A = [7 3 4 1 1 0 0 ;
     2 1 1 5 0 1 0 ;
     1 4 5 2 0 0 1 ]
b = [7; 3; 8]
```

Let's first check the dimension of A and see if it is a full-rank matrix:

```
using LinearAlgebra
m, n = size(A)
@assert rank(A) == m
```

The second line will create an error if rank(A) is not same as m.

We then initialize opt_x and obj:

```
obj = Inf
opt_x = zeros(n)
```

where obj is a scalar variable and opt_x is an n-dimensional array.

To consider all possible combinations of columns in A, we use the Combinatorics[2] package:

---

[2] https://github.com/JuliaMath/Combinatorics.jl

5.2. Searching All Basic Feasible Solutions

```
using Combinatorics
combinations(1:n, m)
```

For example, with $n = 7$ and $m = 3$, we obtain

```
julia> combinations(1:7, 3)
Combinatorics.Combinations{UnitRange{Int64}}(1:7, 3)

julia> collect( combinations(1:7, 3) )
35-element Array{Array{Int64,1},1}:
 [1,2,3]
 [1,2,4]
 [1,2,5]
 [1,2,6]
 [1,2,7]
 [1,3,4]
 [1,3,5]
 ...
 ...
 ...
 [3,5,7]
 [3,6,7]
 [4,5,6]
 [4,5,7]
 [4,6,7]
 [5,6,7]
```

The function `combinations()` basically generates an Iterator object, for which we can retrieve the values using the function `collect()`. Each combination represents the indices of columns in `A`, which corresponds to the indices of basic variables. We will construct a `for`-loop:

```
for b_idx in combinations(1:n, m)
    # see if combination b_idx implies a feasible solution (BFS)
end
```

For each `b_idx`, we construct the basis matrix $\mathbf{B}$, the cost vector for basic variables $\mathbf{c}_B$, and the current value of basic variables $\mathbf{x}_B = \bar{b} = \mathbf{B}^{-1}\mathbf{b}$ as follows:

```
B = A[:, b_idx]
c_B = c[b_idx]
x_B = inv(B) * b
```

where we used `inv(B)` for computing the inverse of **B**. In the first line `A[:, b_idx]` basically means a collection of the columns in `A`. For example `A[:,2]` means the second column, and `A[:,5]` means the fifth column of `A`. If `b_idx=[2,5,6]`, `A[:, b_idx]` will return the second, fifth, and sixth columns of `A` as a matrix:

```
julia> A = [7 3 4 1 1 0 0 ;
            2 1 1 5 0 1 0 ;
            1 4 5 2 0 0 1 ]

julia> b_idx = [2,5,6]

julia> A[:, b_idx]
3x3 Array{Int64,2}:
 3  1  0
 1  0  1
 4  0  0
```

Similarly

```
julia> c = [-3; -2; -1; -5; 0; 0; 0]

julia> c[b_idx]
3-element Array{Int64,1}:
 -2
  0
  0
```

As a next step, we need to check if the current basis implies a feasible solution; i.e. $x_B \geq 0$. For this check, we create a function:

```
function is_nonnegative(x::Vector)
   return length( x[ x .< 0] ) == 0
end
```

5.2. Searching All Basic Feasible Solutions

In the above function definition, `x::Vector` means that this function only accepts a vector type. Note that `x.<0` is an element-wise comparison with 0; therefore the statement `x[x.<0]` returns the elements of `x` that is less than 0. If the length of `x[x.< 0]` is zero, then there is no negative element in `x`, which means vector `x` is nonnegative. To clearly understand what `x[x.<0]` means, see the following examples: a case with some negative elements

```
julia> x = [-1; 2; -2]

julia> x .< 0
3-element BitArray{1}:
  true
 false
  true

julia> x[x .< 0]
2-element Array{Int64,1}:
 -1
 -2

julia> length(x[x .< 0]) == 0
false
```

and a case of all nonnegative elements

```
julia> x = [2; 1; 0]
3-element Array{Int64,1}:
 2
 1
 0

julia> x[x .< 0]
0-element Array{Int64,1}

julia> length(x[x .< 0]) == 0
true
```

If we have a nonnegative x_B, then we compare its objective function value with the smallest objective function value that is stored in `obj`.

```
if is_nonnegative(x_B)
  z = dot(c_B, x_B)
  if z < obj
    obj = z
    opt_x = zeros(n)
    opt_x[b_idx] = x_B
  end
end
```

where we used dot(c_B, x_B) to compute $c_B^\top x_B$.

The complete code is shown below:

Listing 5.1: Searching all BFS *code/chap5/search_bfs.jl*

```
using LinearAlgebra, Combinatorics

function is_nonnegative(x::Vector)
  return length( x[ x .< 0] ) == 0
end

function search_BFS(c, A, b)
  m, n = size(A)
  @assert rank(A) == m

  opt_x = zeros(n)
  obj = Inf

  for b_idx in combinations(1:n, m)
    B = A[:, b_idx]
    c_B = c[b_idx]
    x_B = inv(B) * b

    if is_nonnegative(x_B)
      z = dot(c_B, x_B)
      if z < obj
        obj = z
        opt_x = zeros(n)
        opt_x[b_idx] = x_B
      end
    end
```

5.3. Using the JuMP Package

```
    println("Basis:", b_idx)
    println("\t x_B = ", x_B)
    println("\t nonnegative? ", is_nonnegative(x_B))
    if is_nonnegative(x_B)
      println("\t obj = ", dot(c_B, x_B))
    end

  end

  return opt_x, obj
end
```

For the given problem, the following codes will produce an optimal solution `opt_x` whose objective function value `obj` equals to -5.051724137931034.

```
julia> include("search_bfs.jl")

julia> opt_x, obj = search_BFS(c, A, b)
```

5.3 Using the JuMP Package

Of course, we can also use the JuMP Package to solve the LP problem. While we use JuMP as an interface, we use GLPK as a solver. We can use any other LP solvers installed in your system, for example `Clp`, `Gurobi` or `CPLEX`.

```
using JuMP, GLPK, LinearAlgebra
m = Model(GLPK.Optimizer)
@variable(m, x[1:length(c)] >=0 )
@objective(m, Min, dot(c, x))
@constraint(m, A*x .== b)

JuMP.optimize!(m)
obj0 = JuMP.objective_value(m)
opt_x0 = JuMP.value.(x)
```

In the above, we used `A*x .== b` to define each row as a constraint.

We compare the optimal objective function value obtained by JuMP,

$$-5.051724137931035,$$

with the solution obtained by searching all BFS in the previous section, which was

$$-5.051724137931034.$$

The two solutions are "same" within acceptable precision.

5.4 Pivoting in Tableau Form

In this section, we review pivoting of the Simplex Method in tableau form. We will again consider the following problem instance:

```
c = [-3; -2; -1; -5; 0; 0; 0]
A = [7 3 4 1 1 0 0 ;
     2 1 1 5 0 1 0 ;
     1 4 5 2 0 0 1 ]
b = [7; 3; 8]
```

which may be written in the following tableau form:

	z	x_1	x_2	x_3	x_4	x_5	x_6	x_7	RHS
z	1	3	2	1	5	0	0	0	0
x_5	0	7	3	4	1	1	0	0	7
x_6	0	2	1	1	5	0	1	0	3
x_7	0	1	4	5	2	0	0	1	8

In the above tableau, x_5, x_6, and x_7 are the basic variables, and the corresponding basis $\mathbf{B} = [\mathbf{a}_5\ \mathbf{a}_6\ \mathbf{a}_6]$, which is an identity matrix in this case. The information listed in the above can be described as follows:

	z	x_1	x_2	x_3	x_4	x_5	x_6	x_7	RHS
z	1			$\cdots$ $z_j - c_j$ $\cdots$					$\mathbf{c}_B^\top \mathbf{x}_B$
x_5	0	|	|	|	|	|	|	|	
x_6	0	$\mathbf{y}_1$	$\mathbf{y}_2$	$\mathbf{y}_3$	$\mathbf{y}_4$	$\mathbf{y}_5$	$\mathbf{y}_6$	$\mathbf{y}_7$	$\bar{b}$
x_7	0	|	|	|	|	|	|	|	

5.4. Pivoting in Tableau Form

or

	z	x_1	x_2	x_3	x_4	x_5	x_6	x_7	RHS
z	1	\multicolumn{7}{c	}{$c_B^\top B^{-1} A - c^\top$}	$c_B^\top B^{-1} b$
x_5	0								
x_6	0	\multicolumn{7}{c	}{$Y = B^{-1} A$}	$B^{-1} b$
x_7	0								

Note that $A = [B \ N]$. For basic variables x_B, we have $c_B^\top B^{-1} B - c_B^\top = 0$; therefore $z_j - c_j$ for $j \notin J$. For nonbasic variables x_N, we have $c_B^\top B^{-1} N - c_N^\top$; therefore $z_j - c_j = c_B^\top B^{-1} a_j - c_j$ for $j \in J$.

Coming back to the first numeric tableau, we determine an entering variable. In the current tableau, all nonbasic variables have $z_j - c_j > 0$; we choose the first one, x_1 as the entering variable. After the minimum ratio test, we determine x_5 as the exiting variable, since

$$\left\{\frac{7}{7}, \frac{3}{2}, \frac{8}{1}\right\} = \frac{7}{7}.$$

Pivoting is done in the two steps:

1. To make y_{11} equals to 1, by dividing the row of x_5 by the current value of y_{11}.

2. By elementary row operations using the new row of x_5, to make the first column equals to zero in the rows of z, x_6, and x_7, and to let x_1 replaces x_5 in the basic variables.

That is, after the first step, the tableau looks like:

	z	x_1	x_2	x_3	x_4	x_5	x_6	x_7	RHS
z	1	3	2	1	5	0	0	0	0
x_5	0	**1**	3/7	4/7	1/7	1/7	0	0	1
x_6	0	2	1	1	5	0	1	0	3
x_7	0	1	4	5	2	0	0	1	8

A change is made to the bold-faced numbers. After the second step, the tableau looks like:

	z	x_1	x_2	x_3	x_4	x_5	x_6	x_7	RHS
z	1	0	5/7	-5/7	32/7	-3/7	0	0	-3
x_1	0	1	3/7	4/7	1/7	1/7	0	0	1
x_6	0	0	1/7	-1/7	33/7	-2/7	1	0	1
x_7	0	0	23/7	31/7	13/7	-1/7	0	1	7

A change is made to the bold-faced numbers. Note that in the column of x_1 there is only one non-zero element, which is 1 for the row of x_1.

In our implementation of the Simplex Method, we will follow these elementary row operations.

5.5 Implementing the Simplex Method

In this section, we implement the Simplex Method. Well, we don't intend to create a practically useful code. Instead, we will take this opportunity to learn how one can "translate" the computing works done by hands into Julia codes. We'll try to mimic the pivoting in tableau form as explained in the previous section. Along the way, we'll also learn some advanced features of the Julia language.

To store the information of each tableau, we will create a custom data type:

```
mutable struct SimplexTableau
    z_c      :: Array{Float64}
    Y        :: Array{Float64}
    x_B      :: Array{Float64}
    obj      :: Float64
    b_idx    :: Array{Int64}
end
```

There are five *fields* in this data type. We will use

- z_c for storing the row representing $z_j - c_j$,
- Y for $\mathbf{Y}$,
- x_B for $\mathbf{x}_B = \bar{b} = \mathbf{B}^{-1}\mathbf{b}$,
- obj for $z = \mathbf{c}_B^\top \mathbf{B}^{-1}\mathbf{b}$, and
- b_idx for the indices of basic variables.

The type of each field is also described in the definition of the new data type. After defining it, one may create an instance of `SimplexTableau` that represents the very first tableau in the previous section as follows:

5.5. Implementing the Simplex Method

```
z_c = [3 2 1 5 0 0 0]
Y = [7 3 4 1 1 0 0 ;
     2 1 1 5 0 1 0 ;
     1 4 5 2 0 0 1 ]
x_B = [7; 3; 8]
obj = 0
b_idx = [5, 6, 7]
tableau = SimplexTableau(z_c, Y, x_B, obj, b_idx)
```

Note that `z_c` meant to be a *row* vector. Compare the above code with the tableau form in the previous section. Now `tableau` is a variable of type `SimplexTableau`:

```
julia> typeof(tableau)
SimplexTableau

julia> fieldnames(typeof(tableau))
(:z_c, :Y, :x_B, :obj, :b_idx)
```

We can access the `b_idx` field as follows:

```
julia> tableau.b_idx
3-element Array{Int64,1}:
 5
 6
 7
```

With the new data type defined for representing the tableau, we create a function for the Simplex Method:

```
function simplex_method(c, A, b)
  tableau = initialize(c, A, b)

  while !is_optimal(tableau)
    pivoting!(tableau)
  end

  # compute opt_x (x*) from tableau
```

```
    opt_x = zeros(length(c))
    opt_x[tableau.b_idx] = tableau.x_B

    return opt_x, tableau.obj
end
```

The function `initialize()` first finds an initial BFS and create the first tableau, which is saved in a variable `tableau`. The function `isOptimal()` checks if the current `tableau` is optimal or not. If it is not optimal yet, it performs `pivoting!`. When it is optimal, we terminate the `while` loop and returns an optimal solution with its optimal objective function value. We simple need to prepare three functions:

1. `initialize(c, A, b)`

2. `is_optimal(tableau)`

3. `pivoting!(tableau)`

5.5.1 initialize(c, A, b)

To begin the Simplex Method, we need to have a BFS on our hand. This function `initialize()` will find a BFS and creates the first tableau. The first thing to do in this function is to make it sure that all inputs c, A, and b are arrays of `Float64`:

```
c = Array{Float64}(c)
A = Array{Float64}(A)
b = Array{Float64}(b)
```

This will convert c, A, and b to of type `ArrayFloat64`, if they are not already. We measure the size of A:

```
m, n = size(A)
```

To find an initial BFS, we will create another function `initial_BFS()`:

5.5. Implementing the Simplex Method

```
b_idx, x_B, B = initial_BFS(A,b)
```

Ideally, one uses the two-phase method to find an initial BFS. In this code, for the simplicity, we just search for a possible combination of basis until we find a BFS.

```
function initial_BFS(A, b)
  m, n = size(A)

  comb = collect(combinations(1:n, m))
  for i in length(comb):-1:1
    b_idx = comb[i]
    B = A[:, b_idx]
    x_B = inv(B) * b
    if is_nonnegative(x_B)
      return b_idx, x_B, B
    end
  end

  error("Infeasible")
end
```

The above function tries the last m columns, hoping that they are slack variables. The `is_nonnegative()` function is same as before. Certainly, the `initial_BFS()` function needs improvements. Let us, however, just use the above simple code for now.

Once we obtain `b_idx`, `x_B`, and B from `initial_BFS()`, we compute all necessary values:

```
Y = inv(B) * A
c_B = c[b_idx]
obj = dot(c_B, x_B)

# z_c is a row vector
z_c = zeros(1,n)
n_idx = setdiff(1:n, b_idx)
z_c[n_idx] = c_B' * inv(B) * A[:,n_idx] - c[n_idx]'
```

In the above code, n_idx represents J, the current set of indices for non-basic variables; therefore A[:,n_idx] and c[n_idx] represent $\mathbf{N}$ and $\mathbf{c}_N$, respectively. Finally, create an object of SimplexTableau type, and return it:

```
return SimplexTableau(z_c, Y, x_B, obj, b_idx)
```

The complete initialize() code is shown below:

```
function initialize(c, A, b)
  c = Array{Float64}(c)
  A = Array{Float64}(A)
  b = Array{Float64}(b)

  m, n = size(A)

  # Finding an initial BFS
  b_idx, x_B, B = initial_BFS(A,b)

  Y = inv(B) * A
  c_B = c[b_idx]
  obj = dot(c_B, x_B)

  # z_c is a row vector
  z_c = zeros(1,n)
  n_idx = setdiff(1:n, b_idx)
  z_c[n_idx] = c_B' * inv(B) * A[:,n_idx] - c[n_idx]'

  return SimplexTableau(z_c, Y, x_B, obj, b_idx)
end
```

5.5.2   is_optimal(tableau)

Checking the current tableau is optimal or not is simple: We examine the z_c field of tableau has any positive element. We use the following function:

```
function is_optimal(t::SimplexTableau)
  return is_nonpositive(t.z_c)
end
```

5.5. Implementing the Simplex Method

```
function is_nonpositive(z::Array)
   return length( z[ z .> 0] ) == 0
end
```

where `is_nonpositive()` is defined as similar as `is_nonnegative()`.

5.5.3   pivoting!(tableau)

This function name has an exclamation mark !, which is a coding convention in Julia. The exclamation mark implies that the function changes the content of the input argument; so, warning. Pivoting will change the values in the current tableau by elementary row operations. Instead of creating new objective, we'll be simply changing the values stored in the current `tableau`.

The function looks like:

```
function pivoting!(t::SimplexTableau)
   entering, exiting = pivot_point(t)

   # Perform pivoting
end
```

where we first determine the entering variable and the exiting variable using the `pivot_point()` function. Note that the current tableau information is stored in `t`.

The function `pivot_point()` looks like:

```
function pivot_point(t::SimplexTableau)
   # Finding the entering variable index
   entering = ...

   # min ratio test / finding the exiting variable index
   exiting = ...

   return entering, exiting
end
```

Finding the entering variable can be easily done by searching $z_j - c_j > 0$.

```
entering = findfirst(t.z_c .> 0)[2]
```

where we select the smallest index with $z_j - c_j > 0$. Since `t.z_c` is a row vector, `findfirst()` will return $(1, k)$ if the smallest index is k. Therefore, we need to put [2] at the end to retrieve k only.

Recall that the min-ratio test is done by:

$$x_k = \frac{\bar{b}_r}{y_{rk}} = \min_{i=1,\ldots,m} \left\{ \frac{\bar{b}_i}{y_{ik}} : y_{ik} > 0 \right\}$$

To perform the min-ratio test, we first find rows with $y_{ik} > 0$; that is,

```
pos_idx = findall( t.Y[:, entering] .> 0 )
```

which stores the indices of rows in `t.Y` with strictly positive values. We note that if all elements of `t.Y` are nonnegative, the problem is unbounded; hence we create an error if that is the case:

```
if length(pos_idx) == 0
   error("Unbounded")
end
```

To determine the exiting variable, we use the following code:

```
exiting = pos_idx[ argmin( t.x_B[pos_idx] ./ t.Y[pos_idx, entering] ) ]
```

where `t.Y[pos_idx, entering]` corresponds to y_{ik} with k is same as `entering`, and `t.x_B[pos_idx]` corresponds to $\bar{b}_i$. Note that the function `argmin()` returns the index of `pos_idx` for the minimum ratio. Finally, `pos_idx[argmin(...)]` retrieves the row index of the exiting variable in the tableau, or the index within $\mathbf{x}_B$.

The complete code for `pivot_point` is shown below:

5.5. Implementing the Simplex Method

```
function pivot_point(t::SimplexTableau)
  # Finding the entering variable index
  entering = findfirst(t.z_c .> 0)[2]
  if entering == 0
    error("Optimal")
  end

  # min ratio test / finding the exiting variable index
  pos_idx = findall( t.Y[:, entering] .> 0 )
  if length(pos_idx) == 0
    error("Unbounded")
  end
  exiting = pos_idx[ argmin( t.x_B[pos_idx] ./ t.Y[pos_idx, entering] ) ]

  return entering, exiting
end
```

Note that `entering` is the index in **x**, while `exiting` is the index in $\mathbf{x}_B$. In the tableau, `entering` is the column number, and `exiting` is the row number for pivoting.

Once we have determined `entering` and `exiting`, our first job is to make the value of the pivot point equal to 1.

```
coef = t.Y[exiting, entering]
t.Y[exiting, :] /= coef
t.x_B[exiting] /= coef
```

Note that `a /= b` means `a = a / b`. Similarly, `a += b` means `a = a + b`, and the same for `a -= b` and `a *= b`. The current value in the pivot point is stored in `coef`. The second and third lines are shorthands for

```
t.Y[exiting, :] = t.Y[exiting, :] / coef
t.x_B[exiting] = t.x_B[exiting] / coef
```

All numbers in the exiting row are divided by `coef`. Note that we are changing the contents of the input argument `t` here; that's why we put an exclamation mark in the function name as in `pivoting!()`.

For all other rows in t.Y, we perform row operations to make the values in column entering equal to zero.

```
for i in setdiff(1:m, exiting)
  coef = t.Y[i, entering]
  t.Y[i, :] -= coef * t.Y[exiting, :]
  t.x_B[i] -= coef * t.x_B[exiting]
end
```

Using the row exiting, we update row i to make t.Y[i, entering] zero. We apply the same kind of row operation to the z_c row representing $z_j - c_j$ and the current objective function value obj.

```
coef = t.z_c[entering]
t.z_c -= coef * t.Y[exiting, :]'
t.obj -= coef * t.x_B[exiting]
```

Note that we used t.Y[exiting, :]' with the transpose operator ' at the end. Starting from Julia v0.5, t.Y[exiting, :] returns a column vector, while we want a row vector. One may also use t.Y[exiting:exiting, :] to obtain a row vector.

Finally, we update b_idx that represents the indices of basic variables.

```
t.b_idx[ findall(t.b_idx .== t.b_idx[exiting]) ] = entering
```

Note first that the index of the exiting variable is t.b_idx[exiting]. Among the elements of t.b_idx, find the element with the value t.b_idx[exiting], which is replaced by entering. The above replacement can alternatively be done by the following code:

```
for i in 1:length(t.b_idx)
  if t.b_idx[i] == t.b_idx[exiting]
    t.b_idx[i] = entering
  end
end
```

The complete code is shown below:

5.5. Implementing the Simplex Method

```
function pivoting!(t::SimplexTableau)
  m, n = size(t.Y)

  entering, exiting = pivot_point(t)
  # Pivoting: exiting-row, entering-column
  # updating exiting-row
  coef = t.Y[exiting, entering]
  t.Y[exiting, :] /= coef
  t.x_B[exiting] /= coef

  # updating other rows of Y
  for i in setdiff(1:m, exiting)
    coef = t.Y[i, entering]
    t.Y[i, :] -= coef * t.Y[exiting, :]
    t.x_B[i] -= coef * t.x_B[exiting]
  end

  # updating the row for the reduced costs
  coef = t.z_c[entering]
  t.z_c -= coef * t.Y[exiting, :]
  t.obj -= coef * t.x_B[exiting]

  # Updating b_idx
  t.b_idx[ findall(t.b_idx .== t.b_idx[exiting]) ] = entering
end
```

5.5.4 Creating a Module

To use the simplex method code conveniently, we may create a module for it. Check out the explanation of modules in the official documentation[3].

Our module, named `SimplexMethod`, will look like:

```
module SimplexMethod

  export simplex_method

  type SimplexTableau
```

---
[3]https://docs.julialang.org/en/v1/manual/modules/

```
    ....
    end

    function simplex_method(c, A, b)
        ...
    end

    # all other functions come here
end
```

We want only the `simplex_method()` function is accessible from the outside; therefore `export simplex_method`. Once the module is created and saved in `simplex_method.jl`, one may use the module as follows:

```
include("simplex_method.jl")
using Main.SimplexMethod
simplex_method(c, A, b)
```

We can use our module as if it is a package. Indeed, a package is a module for which Julia knows its saved location.

The entire code, where I added some printing functions, is shown below:

Listing 5.2: Searching all BFS *code/chap5/simplex_method.jl*

```
module SimplexMethod

    using LinearAlgebra, Combinatorics, Printf

    export simplex_method

    mutable struct SimplexTableau
        z_c     ::Array{Float64} # z_j - c_j
        Y       ::Array{Float64} # inv(B) * A
        x_B     ::Array{Float64} # inv(B) * b
        obj     ::Float64        # c_B * x_B
        b_idx   ::Array{Int64}   # indices for basic variables x_B
    end

    function is_nonnegative(x::Vector)
```

5.5. Implementing the Simplex Method

```julia
    return length( x[ x .< 0] ) == 0
end

function is_nonpositive(z::Array)
    return length( z[ z .> 0] ) == 0
end

function initial_BFS(A, b)
  m, n = size(A)

  comb = collect(combinations(1:n, m))
  for i in length(comb):-1:1
    b_idx = comb[i]
    B = A[:, b_idx]
    x_B = inv(B) * b
    if is_nonnegative(x_B)
      return b_idx, x_B, B
    end
  end

  error("Infeasible")
end

function print_tableau(t::SimplexTableau)
  m, n = size(t.Y)

  hline0 = repeat("-", 6)
  hline1 = repeat("-", 7*n)
  hline2 = repeat("-", 7)
  hline = join([hline0, "+", hline1, "+", hline2])

  println(hline)

  @printf("%6s|", "")
  for j in 1:length(t.z_c)
    @printf("%6.2f ", t.z_c[j])
  end
  @printf("| %6.2f\n", t.obj)

  println(hline)

  for i in 1:m
    @printf("x[%2d] |", t.b_idx[i])
```

```julia
      for j in 1:n
        @printf("%6.2f ", t.Y[i,j])
      end
      @printf("| %6.2f\n", t.x_B[i])
    end

    println(hline)
end

function pivoting!(t::SimplexTableau)
  m, n = size(t.Y)

  entering, exiting = pivot_point(t)
  println("Pivoting: entering = x_$entering, exiting = x_$(t.b_idx[exiting])")

  # Pivoting: exiting-row, entering-column
  # updating exiting-row
  coef = t.Y[exiting, entering]
  t.Y[exiting, :] /= coef
  t.x_B[exiting] /= coef

  # updating other rows of Y
  for i in setdiff(1:m, exiting)
    coef = t.Y[i, entering]
    t.Y[i, :] -= coef * t.Y[exiting, :]
    t.x_B[i] -= coef * t.x_B[exiting]
  end

  # updating the row for the reduced costs
  coef = t.z_c[entering]
  t.z_c -= coef * t.Y[exiting, :]'
  t.obj -= coef * t.x_B[exiting]

  # Updating b_idx
  t.b_idx[ findfirst(t.b_idx .== t.b_idx[exiting]) ] = entering
end

function pivot_point(t::SimplexTableau)
  # Finding the entering variable index
  entering = findfirst( t.z_c .> 0)[2]
  if entering == 0
    error("Optimal")
  end
```

5.5. Implementing the Simplex Method

```julia
  # min ratio test / finding the exiting variable index
  pos_idx = findall( t.Y[:, entering] .> 0 )
  if length(pos_idx) == 0
    error("Unbounded")
  end
  exiting = pos_idx[ argmin( t.x_B[pos_idx] ./ t.Y[pos_idx, entering] ) ]

  return entering, exiting
end

function initialize(c, A, b)
  c = Array{Float64}(c)
  A = Array{Float64}(A)
  b = Array{Float64}(b)

  m, n = size(A)

  # Finding an initial BFS
  b_idx, x_B, B = initial_BFS(A,b)

  Y = inv(B) * A
  c_B = c[b_idx]
  obj = dot(c_B, x_B)

  # z_c is a row vector
  z_c = zeros(1,n)
  n_idx = setdiff(1:n, b_idx)
  z_c[n_idx] = c_B' * inv(B) * A[:,n_idx] - c[n_idx]'

  return SimplexTableau(z_c, Y, x_B, obj, b_idx)
end

function is_optimal(t::SimplexTableau)
  return is_nonpositive(t.z_c)
end

function simplex_method(c, A, b)
  tableau = initialize(c, A, b)
  print_tableau(tableau)

  while !is_optimal(tableau)
    pivoting!(tableau)
```

```
    print_tableau(tableau)
  end

  opt_x = zeros(length(c))
  opt_x[tableau.b_idx] = tableau.x_B

  return opt_x, tableau.obj
  end

end
```

A test is run as follows:

```
c = [-3; -2; -1; -5; 0; 0; 0]
A = [7 3 4 1 1 0 0 ;
     2 1 1 5 0 1 0 ;
     1 4 5 2 0 0 1 ]
b = [7; 3; 8]
include("simplex_method.jl")
using Main.SimplexMethod
simplex_method(c, A, b)
```

and the following result is obtained:

```
------+----------------------------------------------+-------
      |   3.00   2.00   1.00   5.00   0.00   0.00   0.00  |   0.00
------+----------------------------------------------+-------
x[ 5] |   7.00   3.00   4.00   1.00   1.00   0.00   0.00  |   7.00
x[ 6] |   2.00   1.00   1.00   5.00   0.00   1.00   0.00  |   3.00
x[ 7] |   1.00   4.00   5.00   2.00   0.00   0.00   1.00  |   8.00
------+----------------------------------------------+-------
Pivoting: entering = x_1, exiting = x_5
------+----------------------------------------------+-------
      |   0.00   0.71  -0.71   4.57  -0.43   0.00   0.00  |  -3.00
------+----------------------------------------------+-------
x[ 1] |   1.00   0.43   0.57   0.14   0.14   0.00   0.00  |   1.00
x[ 6] |   0.00   0.14  -0.14   4.71  -0.29   1.00   0.00  |   1.00
x[ 7] |   0.00   3.57   4.43   1.86  -0.14   0.00   1.00  |   7.00
------+----------------------------------------------+-------
Pivoting: entering = x_2, exiting = x_7
```

5.6. Next Steps

```
------+--------------------------------------------+-------
      |   0.00   0.00  -1.60   4.20  -0.40   0.00  -0.20 |  -4.40
------+--------------------------------------------+-------
x[ 1] |   1.00   0.00   0.04  -0.08   0.16   0.00  -0.12 |   0.16
x[ 6] |   0.00   0.00  -0.32   4.64  -0.28   1.00  -0.04 |   0.72
x[ 2] |   0.00   1.00   1.24   0.52  -0.04   0.00   0.28 |   1.96
------+--------------------------------------------+-------
Pivoting: entering = x_4, exiting = x_6
------+--------------------------------------------+-------
      |   0.00   0.00  -1.31   0.00  -0.15  -0.91  -0.16 |  -5.05
------+--------------------------------------------+-------
x[ 1] |   1.00   0.00   0.03   0.00   0.16   0.02  -0.12 |   0.17
x[ 4] |   0.00   0.00  -0.07   1.00  -0.06   0.22  -0.01 |   0.16
x[ 2] |   0.00   1.00   1.28   0.00  -0.01  -0.11   0.28 |   1.88
------+--------------------------------------------+-------
([0.172414, 1.87931, 0.0, 0.155172, 0.0, 0.0, 0.0], -5.051724137931035)
```

In the last tableau, all $z_j - c_j$ are negative or zero; hence optimal.

5.6 Next Steps

The code that we created is complete by no means. To create a complete, working code for the Simplex Method, we should at least address the following issues:

- To handles inequality constraints by adding slack and surplus variables appropriately (improve `initialize`),

- To find an initial BFS using the two-phase method (improve `initial_BFS`),

- To avoid cycling using Bland's rule or other methods (improve `pivot_point`), and

- To consider lower and upper bounds on the decision variables (improve `pivot_point` and `pivoting!`).

Well, first of all, for the efficiency of the algorithm, it will be better to use the Revised Simplex Method, which reduces the memory usage and the computational efforts required for pivoting.

I leave all these works to the readers. It will be an excellent opportunity for coding practice.

6 Network Optimization Problems

In this section, I will briefly introduce some network optimization problems that are commonly studied and used in operation research and how we can code them in Julia.

6.1 The Minimal-Cost Network-Flow Problem

The minimal-cost network-flow problem deals with a single commodity that need to be distributed over a network. Consider a directed graph $G = (\mathcal{N}, \mathcal{A})$ with the set of nodes $\mathcal{N}$ and the set of links $\mathcal{A}$. The cost of sending a unit flow on link $(i,j) \in \mathcal{A}$ is c_{ij}. There are three types of nodes: source nodes, sink nodes, and intermediate nodes. The single commodity are supplied from source nodes and need to be delivered to sink nodes. Intermediate nodes are neither source nor sink nodes; the commodity just passes through intermediate nodes.

We first define b_i for each node $i \in \mathcal{N}$. For source nodes, b_i denotes the amount of supply and $b_i > 0$. On the other hand, $b_i < 0$ denotes the demand size for sink nodes. For intermediate nodes, $b_i = 0$. For the simplicity, we assume $\sum_{i \in \mathcal{N}} b_i = 0$, which can be relaxed easily.

We also introduce decision variables x_{ij}, which is the amount of flow on link (i,j). We only consider nonnegative flow, i.e., $x_{ij} \geq 0$ for all $(i,j) \in \mathcal{A}$. There may be upper bounds on x_{ij}, denoted by u_{ij}; that is, $x_{ij} \leq u_{ij}$. If there is no upper bound

6.1. The Minimal-Cost Network-Flow Problem

on link (i, j), the constant u_{ij} shall be set to infinity.

The purpose of the minimal-cost network-flow problem is to determine how we should distribute the given supply from the source nodes to meet the demand requirements of the sink nodes with the minimal cost. That is, we solve the following linear programming (LP) problem:

$$\min_{\mathbf{x}} \sum_{(i,j) \in \mathcal{A}} c_{ij} x_{ij} \qquad (6.1)$$

subject to

$$\sum_{(i,j) \in \mathcal{A}} x_{ij} - \sum_{(j,i) \in \mathcal{A}} x_{ji} = b_i \qquad \forall i \in \mathcal{N} \qquad (6.2)$$

$$0 \leq x_{ij} \leq u_{ij} \qquad \forall (i,j) \in \mathcal{A} \qquad (6.3)$$

where (6.2) is the flow conservation constraint, and (6.3) is the nonnegativity constraint.

Since the minimal-cost network flow problem (6.1)–(6.3) is an LP problem, it can be quite easily solved by various methods and there are many LP solvers available. It is indeed a matter of data preparation and implementation of the model in computer programming language. The data we need to prepare are: the unit-flow link cost c_{ij}, and the amount of supply/demand in each node b_i. In addition to these obvious parameters, we also need to prepare more important data: the network structure itself, which describes how nodes and links are connected.

As an example, let us consider a simple network shown in Figure 6.1. For each directed link $(i, j) \in \mathcal{A}$, let us call i the start node, and j the end node. We prepare data in tabular form:

start node i	end node j	c_{ij}	u_{ij}
1	2	2	∞
1	3	5	∞
2	3	3	∞
3	4	1	∞
3	5	2	1
4	1	0	∞
4	5	2	∞
5	2	4	∞

Chapter 6. Network Optimization Problems

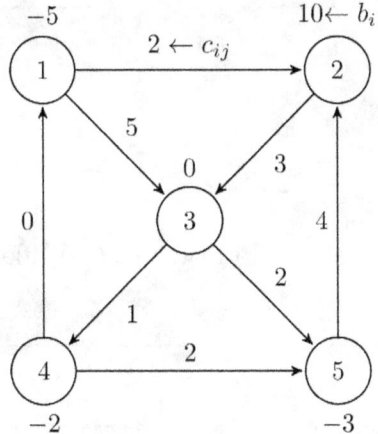

Figure 6.1: A simple network with 5 nodes and 8 links. There is a link with an upper bound on the amount of link flow: $u_{35} = 1$.

Let us use 'Inf' for ∞ in our data preparation. In a spreadsheet, it looks like:

	A	B	C	D	
1	start node i	end node j	c_ij	u_ij	
2	1	2	2	Inf	
3	1	3	5	Inf	
4	2	3	3	Inf	
5	3	4	1	Inf	
6	3	5	2		1
7	4	1	0	Inf	
8	4	5	2	Inf	
9	5	2	4	Inf	
10					

In the CSV format, it looks like:

Listing 6.1: *code/chap6/simple_network.csv*

```
start node i,end node j,c_ij,u_ij
1,2,2,Inf
1,3,5,Inf
```

6.1. The Minimal-Cost Network-Flow Problem

```
2,3,3,Inf
3,4,1,Inf
3,5,2,1
4,1,0,Inf
4,5,2,Inf
5,2,4,Inf
```

As we have seen in Section 3.9, we read this CSV file using **readdlm()** in Julia as follows:

```
using DelimitedFiles
network_data_file = "simple_network.csv"
network_data = readdlm(network_data_file, ',', header=true)
data = network_data[1]
header = network_data[2]
```

Then we save the data in the format we want:

```
start_node = round.(Int64, data[:,1])
end_node = round.(Int64, data[:,2])
c = data[:,3]
u = data[:,4]
```

where **round()** is used to convert, for example, 1.0 to 1. Julia can automatically recognize the text 'Inf' as the numeric data ∞. The vector of upper bound **u** is stored as follows:

```
julia> u
8-element Array{Float64,1}:
  Inf
  Inf
  Inf
  Inf
    1.0
  Inf
  Inf
  Inf
```

Chapter 6. Network Optimization Problems

We see `Inf` of the `Float64` type.

For the vector **b**, we also prepare a CSV file:

	A	B
1	node i	b_i
2	1	-5
3	2	10
4	3	0
5	4	-2
6	5	-3
7		

and read the CSV file in Julia:

```
using DelimitedFiles
network_data2_file = "simple_network_b.csv"
network_data2 = readdlm(network_data2_file, ',', header=true)
data2 = network_data2[1]
hearder2 = network_data2[2]

b = data2[:,2]
```

While doing Julia programming for a network optimization problem, we may need to know the number of nodes and the number of links in the graph. We may explicitly specify those two numbers in a CSV file and read it to Julia. If the numbering of nodes starts from 1 and all positive integers are used without missing any number in the middle, we know that the biggest number used in `start_node` and `end_node` is equal to the number of all nodes in the graph. We have the following code:

```
no_node = max( maximum(start_node), maximum(end_node) )
no_link = length(start_node)
```

where the number of links is simply the number of elements in `start_node`, or `end_node`—they should be same. Note the difference of `max()` and `maximum()`: `max()` is used to compare two different numbers and `maximum()` is used to identify the biggest number among all elements in a vector.

Now we create an array object for the set of nodes $\mathcal{N}$ and the set of links $\mathcal{A}$:

127

6.1. The Minimal-Cost Network-Flow Problem

```
nodes = 1:no_node
links = Tuple( (start_node[i], end_node[i]) for i in 1:no_link )
```

The set of links looks like:

```
julia> links
((1, 2), (1, 3), (2, 3), (3, 4), (3, 5), (4, 1), (4, 5), (5, 2))
```

We will use this array `links` for modeling the minimal-cost network-flow problem. Accordingly, we prepare **c** and **u** in the format of dictionaries:

```
c_dict = Dict(links .=> c)
u_dict = Dict(links .=> u)
```

For example, `c_dict` looks like:

```
julia> c_dict
Dict{Tuple{Int64,Int64},Float64} with 8 entries:
  (3, 5) => 2.0
  (4, 5) => 2.0
  (1, 2) => 2.0
  (2, 3) => 3.0
  (5, 2) => 4.0
  (4, 1) => 0.0
  (1, 3) => 5.0
  (3, 4) => 1.0
```

We are finally ready for writing the minimal-cost network-flow problem using JuMP. First, import the necessary packages:

```
using JuMP, GLPK
```

We prepare an optimization model:

```
mcnf = Model(GLPK.Optimizer)
```

We define the decision variables:

```
@variable(mcnf, 0<= x[link in links] <= u_dict[link])
```

where the bounds on **x** are introduced at the same time. We set the objective:

```
@objective(mcnf, Min, sum(c_dict[link] * x[link] for link in links))
```

Here is the best part. We add the flow conservation constraints:

```
for i in nodes
    @constraint(mcnf, sum(x[(ii,j)] for (ii,j) in links if ii==i)
                    - sum(x[(j,ii)] for (j,ii) in links if ii==i) == b[i])
end
```

where `ii` is a dummy index for `i`. Compare the above code with the mathematical expression:

$$\sum_{(i,j)\in\mathcal{A}} x_{ij} - \sum_{(j,i)\in\mathcal{A}} x_{ji} = b_i \qquad \forall i \in \mathcal{N}$$

The Julia code is almost a direct translation of the original mathematical expression.

When printed, the model looks like:

```
julia> print(mcnf)
Min 2 x[(1, 2)] + 5 x[(1, 3)] + 3 x[(2, 3)] + x[(3, 4)]
    + 2 x[(3, 5)] + 2 x[(4, 5)] + 4 x[(5, 2)]
Subject to
 x[(1, 2)]  0.0
 x[(1, 3)]  0.0
 x[(2, 3)]  0.0
 x[(3, 4)]  0.0
 x[(3, 5)]  0.0
 x[(4, 1)]  0.0
```

6.1. The Minimal-Cost Network-Flow Problem

```
x[(4, 5)]    0.0
x[(5, 2)]    0.0
x[(1, 2)]    Inf
x[(1, 3)]    Inf
x[(2, 3)]    Inf
x[(3, 4)]    Inf
x[(3, 5)]    1.0
x[(4, 1)]    Inf
x[(4, 5)]    Inf
x[(5, 2)]    Inf
x[(1, 2)] + x[(1, 3)] - x[(4, 1)] = -5.0
x[(2, 3)] - x[(1, 2)] - x[(5, 2)] = 10.0
x[(3, 4)] + x[(3, 5)] - x[(1, 3)] - x[(2, 3)] = 0.0
x[(4, 1)] + x[(4, 5)] - x[(3, 4)] = -2.0
x[(5, 2)] - x[(3, 5)] - x[(4, 5)] = -3.0
```

We solve the problem:

```
JuMP.optimize!(mcnf)
```

We retrieve the optimal objective function value and the optimal solution we obtained:

```
obj = JuMP.objective_value(mcnf)
x_star = JuMP.value.(x)
```

We observe that `x_star` is a dictionary type, particularly defined by JuMP:

```
julia> x_star = JuMP.value.(x)
1-dimensional DenseAxisArray{Float64,1,...} with index sets:
    Dimension 1, ((1, 2), (1, 3), (2, 3), (3, 4), (3, 5), (4, 1), (4, 5), (5, 2))
And data, a 8-element Array{Float64,1}:
 0.0
 0.0
 10.0
 9.0
 1.0
 5.0
```

Chapter 6. Network Optimization Problems

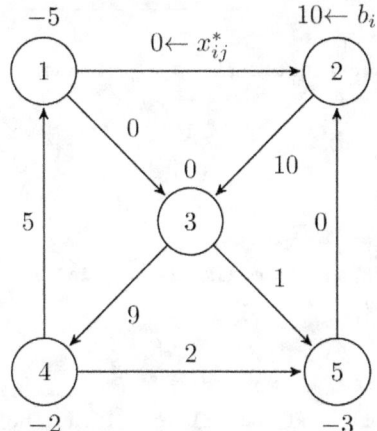

Figure 6.2: The optimal solution of the minimal-cost network-flow problem

```
2.0
0.0
```

This optimal solution is presented in Figure 6.2.
The complete code is presented:

Listing 6.2: Min-Cost Network-Flow Problem
code/chap6/mcnf_example1.jl

```julia
# Importing packages
using JuMP, GLPK, DelimitedFiles

# Data Preparation
network_data_file = "simple_network.csv"
network_data = readdlm(network_data_file, ',', header=true)
data = network_data[1]
header = network_data[2]

start_node = round.(Int64, data[:,1])
end_node = round.(Int64, data[:,2])
c = data[:,3]
u = data[:,4]
```

6.1. The Minimal-Cost Network-Flow Problem

```julia
network_data2_file = "simple_network_b.csv"
network_data2 = readdlm(network_data2_file, ',', header=true)
data2 = network_data2[1]
hearder2 = network_data2[2]

b = data2[:,2]

# number of nodes and number of links
no_node = max( maximum(start_node), maximum(end_node) )
no_link = length(start_node)

# Creating a graph
nodes = 1:no_node
links = Tuple( (start_node[i], end_node[i]) for i in 1:no_link )
c_dict = Dict(links .=> c)
u_dict = Dict(links .=> u)

# Preparing an optimization model
mcnf = Model(GLPK.Optimizer)

# Defining decision variables
@variable(mcnf, 0<= x[link in links] <= u_dict[link])

# Setting the objective
@objective(mcnf, Min, sum( c_dict[link] * x[link] for link in links) )

# Adding the flow conservation constraints
for i in nodes
  @constraint(mcnf, sum(x[(ii,j)] for (ii,j) in links if ii==i )
                  - sum(x[(j,ii)] for (j,ii) in links if ii==i ) == b[i])
end

print(mcnf)
JuMP.optimize!(mcnf)
obj = JuMP.objective_value(mcnf)
x_star = JuMP.value.(x)

println("The optimal objective function value is = $obj")
println(x_star.data)
```

Since the minimal-cost network-flow problem is a general form of many other problems, we save the code as a separate function. We prepare a `mcnf.jl` file and define a function `minimal_cost_network_flow`:

Listing 6.3: Min-Cost Network-Flow Problem
code/chap6/mcnf.jl

```
function minimal_cost_network_flow(nodes, links, c_dict, u_dict, b)
  mcnf = Model(GLPK.Optimizer)

  @variable(mcnf, 0<= x[link in links] <= u_dict[link])

  @objective(mcnf, Min, sum( c_dict[link] * x[link] for link in links) )

  for i in nodes
    @constraint(mcnf, sum(x[(ii,j)] for (ii,j) in links if ii==i )
                    - sum(x[(j,ii)] for (j,ii) in links if ii==i ) == b[i])
  end

  JuMP.optimize!(mcnf)
  obj = JuMP.objective_value(mcnf)
  x_star = JuMP.value.(x)

  return x_star, obj
end
```

After including the function definition by

```
include("mcnf.jl")
```

we can solve the minimal-cost network-flow problem by a function call as follows:

```
x_star, obj =
  minimal_cost_network_flow(nodes, links, c_dict, u_dict, b)
```

6.2 The Transportation Problem

The transportation problem is a special case of the minimal-cost network-flow problem. There are only source and sink nodes without any no intermediate node. Any

6.2. The Transportation Problem

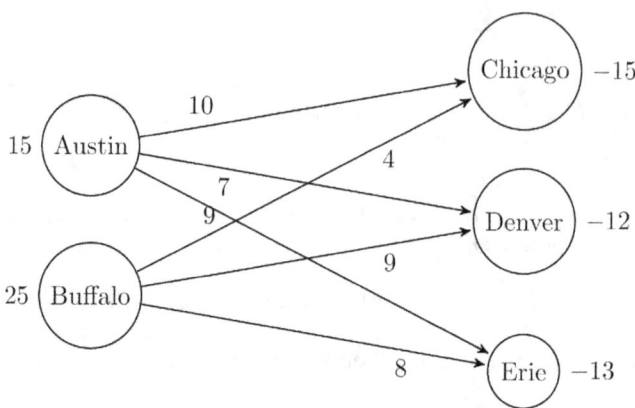

Figure 6.3: A simple example of the transportation problem.

source node is directly connected with all sink nodes, and any sink node is directly connected with all source nodes. For example, see Figure 6.3. Austin and Buffalo are source nodes (or supply nodes), while Chicago, Denver, and Erie are sink nodes (or demand nodes).

Let the set of supply nodes $\mathcal{N}_S$ and the set of demand nodes $\mathcal{N}_D$. We should have $\mathcal{N}_S \cap \mathcal{N}_D = \emptyset$, and $\mathcal{N}_S \cup \mathcal{N}_D = \mathcal{N}$. We can formulate the transportation problem as follows:

$$\min \sum_{i \in \mathcal{N}_S} \sum_{j \in \mathcal{N}_D} c_{ij} x_{ij}$$

subject to

$$\sum_{j \in \mathcal{N}_D} x_{ij} = s_i \quad \forall i \in \mathcal{N}_S$$

$$\sum_{i \in \mathcal{N}_S} x_{ij} = d_j \quad \forall j \in \mathcal{N}_D$$

$$x_{ij} \geq 0 \quad \forall i \in \mathcal{N}_S, j \in \mathcal{N}_D$$

where c_{ij} is the unit transportation cost from supply node i to demand node j, and s_i and d_j are *nonnegative* constants for the supply and demand amounts, respectively. Without loss of generality, we assume $\sum_{i \in \mathcal{N}_S} s_i = \sum_{j \in \mathcal{N}_D} d_j$.

We may have the following tabular data for the problem in Figure 6.3:

Chapter 6. Network Optimization Problems

		15	12	13
		Chicago	Denver	Erie
15	Austin	10	7	9
25	Buffalo	4	9	8

In a spreadsheet, we have:

	A	B	C	D	E
1			15	12	13
2			Chicago	Denver	Erie
3		15 Austin	10	7	9
4		25 Buffalo	4	9	8
5					

The first row is for d_j, while the first column is for s_i. After saving the data in the CSV format, we read the data:

```
julia> using DelimitedFiles
julia> data = readdlm("transportation.csv", ',')
4x5 Array{Any,2}:
  ""    ""           15        12        13
  ""    ""         "Chicago"  "Denver"  "Erie"
  15   "Austin"     10         7         9
  25   "Buffalo"     4         9         8
```

Note that we did not use the header option. The read data looks:
The list of supply nodes and **s** can be accessed by:

```
julia> supply_nodes = data[3:end, 2]
2-element Array{Any,1}:
 "Austin"
 "Buffalo"

julia> s = data[3:end, 1]
2-element Array{Any,1}:
 15
 25
```

Similarly for demand nodes:

135

6.2. The Transportation Problem

```
julia> demand_nodes = data[2, 3:end]
3-element Array{Any,1}:
 "Chicago"
 "Denver"
 "Erie"

julia> d = data[1, 3:end]
3-element Array{Any,1}:
 15
 12
 13
```

Finally the unit transportation cost:

```
julia> c = data[3:end, 3:end]
2x3 Array{Any,2}:
 10  7  9
  4  9  8
```

To make modeling by the JuMP package easier, we prepare dictionary objects for s, d, and c. For the dictionary object `s_dict`, keys are in `supply_nodes` and values are in s; the saved orders are same in both arrays. We can simply do the following:

```
julia> s_dict = Dict(supply_nodes .=> s)
Dict{Any,Any} with 2 entries:
  "Buffalo" => 25
  "Austin"  => 15
```

Similarly, for the demand nodes:

```
julia> d_dict = Dict(demand_nodes .=> d)
Dict{Any,Any} with 3 entries:
  "Chicago" => 15
  "Erie"    => 13
  "Denver"  => 12
```

For c, we can do a double-loop inside `Dict` as follows:

Chapter 6. Network Optimization Problems

```julia
c_dict = Dict( (supply_nodes[i], demand_nodes[j]) => c[i,j]
             for i in 1:length(supply_nodes), j in 1:length(demand_nodes) )
```

which results in:

```julia
julia> c_dict
Dict{Tuple{SubString{String},SubString{String}},Int64} with 6 entries:
  ("Austin", "Chicago")  => 10
  ("Buffalo", "Denver")  => 9
  ("Buffalo", "Chicago") => 4
  ("Austin", "Erie")     => 9
  ("Austin", "Denver")   => 7
  ("Buffalo", "Erie")    => 8
```

Then we are done with the data preparation.

The optimization model can be written as easy as the following code:

```julia
using JuMP, GLPK
tp = Model(GLPK.Optimizer)

@variable(tp, x[supply_nodes, demand_nodes] >= 0)
@objective(tp, Min, sum(c_dict[i,j]*x[i,j]
                        for i in supply_nodes, j in demand_nodes))
for i in supply_nodes
  @constraint(tp, sum(x[i,j] for j in demand_nodes) == s_dict[i] )
end
for j in demand_nodes
  @constraint(tp, sum(x[i,j] for i in supply_nodes) == d_dict[j] )
end
```

The constructed model looks:

```julia
julia> print(tp)
Min 10 x[Austin,Chicago] + 7 x[Austin,Denver] + 9 x[Austin,Erie]
 + 4 x[Buffalo,Chicago] + 9 x[Buffalo,Denver] + 8 x[Buffalo,Erie]
Subject to
 x[Austin,Chicago]  0.0
 x[Austin,Denver]   0.0
```

6.2. The Transportation Problem

```
x[Austin,Erie]   0.0
x[Buffalo,Chicago]   0.0
x[Buffalo,Denver]   0.0
x[Buffalo,Erie]   0.0
x[Austin,Chicago] + x[Austin,Denver] + x[Austin,Erie] = 15.0
x[Buffalo,Chicago] + x[Buffalo,Denver] + x[Buffalo,Erie] = 25.0
```

and its solutions are:

```
julia> JuMP.optimize!(tp)

julia> x_star = JuMP.value.(x)
2-dimensional JuMPArray{Float64,2,...} with index sets:
    Dimension 1, Any["Austin", "Buffalo"]
    Dimension 2, Any["Chicago", "Denver", "Erie"]
And data, a 2×3 Array{Float64,2}:
  0.0  12.0   3.0
 15.0   0.0  10.0

julia> x_star["Austin", "Denver"]
12.0
```

The complete code is presented:

Listing 6.4: Transportation Problem
code/chap6/transportation1.jl

```julia
using JuMP, GLPK, DelimitedFiles

# Reading the data file and preparting arrays
data_file = "transportation.csv"
data = readdlm(data_file, ',')

supply_nodes = data[3:end, 2]
s = data[3:end, 1]

demand_nodes = collect(data[2, 3:end])
d = collect(data[1, 3:end])
```

```
c = data[3:end, 3:end]

# Converting arrays to dictionaries
s_dict = Dict(supply_nodes .=> s)
d_dict = Dict(demand_nodes .=> d)
c_dict = Dict( (supply_nodes[i], demand_nodes[j]) => c[i,j]
         for i in 1:length(supply_nodes), j in 1:length(demand_nodes) )

# Preparing an Optimization Model
tp = Model(GLPK.Optimizer)

@variable(tp, x[supply_nodes, demand_nodes] >= 0)
@objective(tp, Min, sum(c_dict[i,j]*x[i,j]
                    for i in supply_nodes, j in demand_nodes))
for i in supply_nodes
  @constraint(tp, sum(x[i,j] for j in demand_nodes) == s_dict[i] )
end
for j in demand_nodes
  @constraint(tp, sum(x[i,j] for i in supply_nodes) == d_dict[j] )
end

print(tp)
JuMP.optimize!(tp)
obj = JuMP.objective_value(tp)
x_star = JuMP.value.(x)

for s in supply_nodes, d in demand_nodes
    println("from $s to $d: ", x_star[s, d])
end
```

6.3 The Shortest Path Problem

The shortest path problem is perhaps the most famous and important problem in network optimization. It finds a path from a given origin node to a given destination node with the least path cost. It is a special case of the minimal-cost network-flow problem; therefore the shortest-path problem can be solved as a linear programming problem. The problem in linear programming form looks as follows:

$$\min_{\mathbf{x}} \sum_{(i,j) \in \mathcal{A}} c_{ij} x_{ij} \qquad (6.4)$$

6.3. The Shortest Path Problem

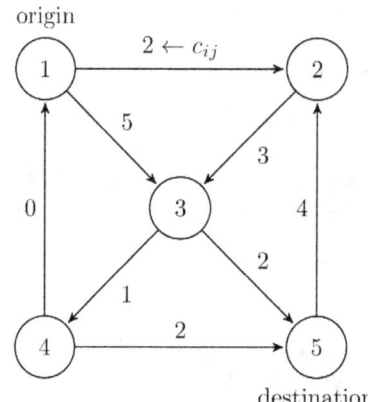

Figure 6.4: The same network as in Figure 6.1, now for the shortest path problem between origin node 1 to destination node 5

subject to

$$\sum_{(i,j)\in \mathcal{A}} x_{ij} - \sum_{(j,i)\in \mathcal{A}} x_{ji} = \begin{cases} 1 & \text{if } i \text{ is the origin node} \\ -1 & \text{if } i \text{ is the destination node} \\ 0 & \text{otherwise} \end{cases} \quad \forall i \in \mathcal{N} \quad (6.5)$$

$$x_{ij} \in \{0, 1\} \quad \forall (i,j) \in \mathcal{A} \quad (6.6)$$

where the binarity constraint (6.6) can be relaxed as $0 \leq x_{ij} \leq 1$ due to the total unimodularity property of the problem structure.

While the shortest path problem can certainly be solved as an LP, there are more efficient algorithms such as Dijkstra's algorithm. The Julia Language provides the `LightGraphs` package for handling graph structures and algorithms for shortest paths. We will figure out how to use functions from the `LightGraphs` package.

First consider the example network in Figure 6.4, which is a representation of Figure 6.1. For preparing and reading data files, please refer to Section 6.1; the only difference is the upper bound u is irrelevant in the shortest path problem.

Chapter 6. Network Optimization Problems

```julia
using DelimitedFiles
network_data_file = "simple_network.csv"
network_data = readdlm(network_data_file, ',', header=true)
data = network_data[1]
header = network_data[2]

start_node = round.(Int64, data[:,1])
end_node = round.(Int64, data[:,2])
c = data[:,3]
```

We have three arrays ready: `start_node`, `end_node`, and `c`. We also set the origin and destination:

```julia
origin = 1
destination = 5
```

and determine the number of nodes and the number of links:

```julia
no_node = max( maximum(start_node), maximum(end_node) )
no_link = length(start_node)
```

Using the `LightGraphs` package, we will create a graph object:

```julia
using LightGraphs
graph = Graph(no_node)
distmx = Inf*ones(no_node, no_node)
```

We just created a graph object, specifying the number of nodes. We will specify how nodes are connected. We also created `distmx` to specify the distance between each nodes. Each pair of nodes has a distance of infinity, meaning disconnected, at this moment. Note that `LightGraphs` requires a distance *matrix*, while we stored the distance data `c` in the vector form.

We add links to the graph object and specify the distance of each link:

141

6.3. The Shortest Path Problem

```
for i in 1:no_link
  add_edge!(graph, start_node[i], end_node[i])
  distmx[start_node[i], end_node[i]] = c[i]
end
```

Note that Julia functions with the exclamation mark '!' changes the value of the argument. In this case, `add_edge!()` changes the inner value of `graph`, since we are adding links one by one. This procedure only construct the organization of the network without specifying how long each link is.

Now, we run Dijkstra's algorithm:

```
state = dijkstra_shortest_paths(graph, origin, distmx)
```

Note that in the above code, `destination` is not passed. It is because Dijkstra's algorithm finds shortest paths from a single origin to *all* nodes in the network. Such algorithms are called one-to-all algorithms.

The results of Dijkstra's algorithm are stored in the `state` variable. To retrieve the results in useful forms, we need some work. Two forms will be most useful: (1) the shortest path as an ordered list of nodes, and (2) a binary vector **x** wherein the i-th component $x_i = 1$ if i-th link is used in the shortest path. In this purpose we use a function from the `LightGraphs` package called, `enumerate_paths()`, and a custom function `getShortestX`. See the full code:

Listing 6.5: Dijkstra's algorithm
code/chap6/ssp_example1.jl

```
# Importing packages
using LightGraphs, DelimitedFiles

# Retrieves 0-1 'x' vector from the 'state'
function getShortestX(state, start_node, end_node, origin, destination)
  _x = zeros(Int, length(start_node))
  _path = enumerate_paths(state, destination)

  for i=1:length(_path)-1
    _start = _path[i]
```

```julia
        _end = _path[i+1]
        
        for j=1:length(start_node)
            if start_node[j]==_start && end_node[j]==_end
                _x[j] = 1
                break
            end
        end
        
    end
    _x
end


# Data Preparation
network_data_file = "simple_network.csv"
network_data = readdlm(network_data_file, ',', header=true)
data = network_data[1]
header = network_data[2]

start_node = round.(Int64, data[:,1])
end_node = round.(Int64, data[:,2])
c = data[:,3]

origin = 1
destination = 5

# number of nodes and number of links
no_node = max( maximum(start_node), maximum(end_node) )
no_link = length(start_node)

# Creating a graph
graph = Graph(no_node)
distmx = Inf*ones(no_node, no_node)

# Adding links to the graph
for i=1:no_link
    add_edge!(graph, start_node[i], end_node[i])
    distmx[start_node[i], end_node[i]] = c[i]
end

# Run Dijkstra's Algorithm from the origin node to all nodes
state = dijkstra_shortest_paths(graph, origin, distmx)
```

6.4. Implementing Dijkstra's Algorithm

```
# Retrieving the shortest path
path = enumerate_paths(state, destination)
println("The shortest path is:", path)

# Retrieving the 'x' variable in a 0-1 vector
x = getShortestX(state, start_node, end_node, origin, destination)
println("x vector:", x)

# The cost of shortest path
println("Cost is $(state.dists[destination])") # directly from state
println("Cost is $(c' * x)")                    # computing from c and x
```

The result looks like:

```
julia> include("ssp_example1.jl")
The shortest path is:[1, 3, 5]
x vector:[0, 1, 0, 0, 1, 0, 0, 0]
Cost is 7.0
Cost is 7.0
```

One may wish to see the x vector with the start and end nodes:

```
julia> [start_node end_node x]
8x3 Array{Int64,2}:
 1  2  0
 1  3  1
 2  3  0
 3  4  0
 3  5  1
 4  1  0
 4  5  0
 5  2  0
```

6.4 Implementing Dijkstra's Algorithm

In the previous section, we have used the `LightGraphs` package to solve the shortest path problem. In this section, we will implement Dijkstra's algorithm by ourself.

Of course, the code will not be as efficient as the `LightGraphs` package. Our own implementation will be a good practice of Julia programming and it may be a good reference for implementing other label-setting or label-correction algorithms.

The origin node is o and the destination node is d. Let's look at the algorithm description:

- **Step 0.** Initialize: $w_1 = 0$ and $\mathcal{X} = \{o\}$.

- **Step 1.** When $\overline{\mathcal{X}} = \mathcal{N} \setminus \mathcal{X}$, find the set

$$(\mathcal{X}, \overline{\mathcal{X}}) = \{(i, j) : i \in \mathcal{X}, j \in \overline{\mathcal{X}}\}$$

- **Step 2.** Find a link (p, q) such that

$$w_p + c_{pq} = \min\left\{w_i + c_{ij} : (i,j) \in (\mathcal{X}, \overline{\mathcal{X}}) \text{ and } (i,j) \in \mathcal{A}\right\}$$

- **Step 3.** Set $w_q = w_p + c_{pq}$ and add the node q to the set $\mathcal{X}$.

- **Step 4.** If the new $\overline{\mathcal{X}} = \mathcal{N} \setminus \mathcal{X}$ is an empty set, stop. Otherwise, go to Step 1 and repeat.

Alternatively, in Step 4, one can terminate the algorithm once the destination node d is labeled, i.e. when $q = d$, when the shortest path to one destination is of interest. For the details of the algorithm see Bazaraa et al. (2011)[1] or Ahuja et al. (1993)[2].

If we look at the algorithms, we basically need the set of nodes $\mathcal{N}$, the set of links $\mathcal{A}$, and c_{ij} as input data. As described in the previous section, we have `start_node`, `end_node`, and `c` for these data. The algorithm will update the values of a vector `w` and the two sets $\mathcal{X}$ and $\overline{\mathcal{X}}$.

Let's first prepare variables. Here is $\mathcal{N}$:

---

[1] Bazaraa, M.S., Jarvis, J.J. and Sherali, H.D., 2011. Linear programming and network flows. John Wiley & Sons.
[2] Ahuja, R.K., Magnanti, T.L. and Orlin, J.B., 1993. Network flows: Theory, algorithms, and applications. Prentice Hall

6.4. Implementing Dijkstra's Algorithm

```
julia> N = Set(1:no_node)
Set([4,2,3,5,1])
```

where `Set` is used. We can think of `Set` as a special type of arrays. This literally represents a set and is useful when the order of elements saved in the array is unimportant. Next, we prepare $\mathcal{A}$ and **c** as a dictionary:

```
links = Tuple( (start_node[i], end_node[i]) for i in 1:no_link )
A = Set(links)
c_dict = Dict(links .=> c)
```

The set **A** looks like:

```
julia> A
Set(Tuple{Int64,Int64}[(3,5), (4,5), (1,2), (2,3), (5,2), (4,1), (1,3), (3,4)])
```

Then, we prepare **w** as follows:

```
w = Array{Float64}(undef, no_node)
```

where we did not assign any value yet. It will have some default values.

In Step 0, we initialize as follows:

```
w[origin] = 0
X = Set{Int}([origin])
Xbar = setdiff(N, X)
```

where we conveniently computed $\overline{\mathcal{X}} = \mathcal{N} \setminus \mathcal{X}$ using `setdiff()`.

For the iterations in Steps 1 to 4, we will use a `while`-loop as follows:

```
while !isempty(Xbar)
    # Step 1
    # Step 2
    # Step 3
    # Step 4
end
```

Chapter 6. Network Optimization Problems

The statement `isempty(Xbar)` will return `true` if there is no element in `Xbar`, i.e. $\overline{\mathcal{X}} = \emptyset$. Therefore, the above `while`-loop will repeat the steps while `Xbar` is not an empty set and terminates when it is empty.

In Step 1, we find the set $(\mathcal{X}, \overline{\mathcal{X}}) = \{(i,j) : i \in \mathcal{X}, j \in \overline{\mathcal{X}}\}$, named as `XX`:

```
XX = Set{Tuple{Int,Int}}()
for i in X, j in Xbar
  if (i,j) in A
    push!(XX, (i,j))
  end
end
```

where the two `for`-loops are used. Whenever we find a link from `X` to `Xbar`, we add it to `XX`.

In Step 2, we find

$$w_p + c_{pq} = \min\left\{w_i + c_{ij} : (i,j) \in (\mathcal{X}, \overline{\mathcal{X}}) \text{ and } (i,j) \in \mathcal{A}\right\}$$

as follows:

```
min_value = Inf
q = 0
for (i,j) in XX
  if w[i] + c_dict[(i,j)] < min_value
    min_value = w[i] + c_dict[(i,j)]
    q = j
  end
end
```

where `min_value` is used to record $w_p + c_{pq}$ and `q` to record q. Note that `min_value` is initialized as ∞ and updated during the search of the minimum.

Step 3 is as simple as:

```
w[q] = min_value
push!(X, q)
```

Step 4 simply updates `Xbar`:

6.4. Implementing Dijkstra's Algorithm

```
Xbar = setdiff(N, X)
```

After this update of `Xbar`, the `while`-loop will evaluate the condition `!isempty(Xbar)` again, to determine to continue or stop. When the `while`-loop stops, we obtain `w[destination]` as the length of the shortest path from origin to destination.

The complete code is provided:

Listing 6.6: An implementation of Dijkstra's algorithm
code/chap6/ssp_example2.jl

```julia
using DelimitedFiles

# Data Preparation
network_data_file = "simple_network.csv"
network_data = readdlm(network_data_file, ',', header=true)
data = network_data[1]
header = network_data[2]

start_node = round.(Int64, data[:,1])
end_node = round.(Int64, data[:,2])
c = data[:,3]

origin = 1
destination = 5

# number of nodes and number of links
no_node = max( maximum(start_node), maximum(end_node) )
no_link = length(start_node)

# Preparing sets and variables
N = Set(1:no_node)
links = Tuple( (start_node[i], end_node[i]) for i in 1:no_link )
A = Set(links)
c_dict = Dict(links .=> c)

function my_dijkstra(N, A, c_dict)
    # Preparing an array
    w = Array{Float64}(undef, no_node)

    # Step 0
```

```
    w[origin] = 0
    X = Set{Int}([origin])
    Xbar = setdiff(N, X)

    # Iterations for Dijkstra's algorithm
    while !isempty(Xbar)
      # Step 1
      XX = Set{Tuple{Int,Int}}()
      for i in X, j in Xbar
        if (i,j) in A
          push!(XX, (i,j))
        end
      end

      # Step 2
      min_value = Inf
      q = 0
      for (i,j) in XX
        if w[i] + c_dict[(i,j)] < min_value
          min_value = w[i] + c_dict[(i,j)]
          q = j
        end
      end

      # Step 3
      w[q] = min_value
      push!(X, q)

      # Step 4
      Xbar = setdiff(N, X)
    end

    return w
end

w = my_dijkstra(N, A, c_dict)

println("The length of node $origin to node $destination is: ", w[destination])
```

The result is:

6.4. Implementing Dijkstra's Algorithm

```
julia> include("ssp_example2.jl")
The length of node 1 to node 5 is: 7.0
```

7

Interior Point Methods

In this chapter, we will see how interior point methods work for solving linear programming problems. Unlike the Simplex Method, which explores the extreme points on the boundary of the feasible set, interior point methods literally explore the strict interior of the feasible set and converge to an optimal solution. Mathematical materials for interior point methods presented in this chapter are a summary of Bertsimas and Tsitsiklis (1997).[1] Two interior points methods will be introduced: the affine scaling algorithm and the primal path following algorithm.

7.1 The Affine Scaling Algorithm

Consider a linear programming problem:

$$\min_{\mathbf{x}} \quad \mathbf{c}^\top \mathbf{x}$$
$$\text{s.t.} \quad \mathbf{A}\mathbf{x} = \mathbf{b}$$
$$\mathbf{x} \geq 0$$

and its dual problem:

$$\max_{\mathbf{p}} \quad \mathbf{p}^\top \mathbf{x}$$

---

[1] Bertsimas, D. and Tsitsiklis, J.N., 1997. Introduction to linear optimization. Belmont, MA: Athena Scientific.

7.1. The Affine Scaling Algorithm

$$\text{s.t.} \quad \mathbf{p}^\top \mathbf{A} \leq \mathbf{c}^\top$$

where $\mathbf{p}$ is the dual variable. We call any $\mathbf{x}$ such that $\mathbf{Ax} = \mathbf{b}$ and $\mathbf{x} > \mathbf{0}$ an interior point.

The affine scaling algorithm is based on ellipsoids defined within the feasible region of the primal problem. For any given strictly positive vector $\mathbf{y} \in \mathbb{R}^n$, suppose $\mathbf{x}$ satisfies:

$$\sum_{i=1}^{n} \frac{(x_i - y_i)^2}{y_i^2} \leq \beta^2$$

for some constant $\beta \in (0, 1)$, or equivalently in a vector form:

$$\left\| \mathbf{Y}^{-1}(\mathbf{x} - \mathbf{y}) \right\| \leq \beta$$

where $\mathbf{Y} = \text{diag}(y_1, ..., y_n)$ is a diagonal matrix. Then, we can show that $\mathbf{x} > \mathbf{0}$. This means that we can remain in the interior of the feasible region, as long as we stay within the ellipsoid around $\mathbf{y}$ and $\mathbf{Ay} = \mathbf{b}$.

Consider the following optimization problem for any given $\mathbf{y}$:

$$\min_{\mathbf{x}} \quad \mathbf{c}^\top \mathbf{x}$$
$$\text{s.t.} \quad \mathbf{Ax} = \mathbf{b}$$
$$\left\| \mathbf{Y}^{-1}(\mathbf{x} - \mathbf{y}) \right\| \leq \beta$$

where the non-negativity has been replaced by the ellipsoidal constraint. If we define a direction $\mathbf{d} = \mathbf{x} - \mathbf{y}$, then we can write an equivalent optimization problem:

$$\min_{\mathbf{d}} \quad \mathbf{c}^\top \mathbf{d}$$
$$\text{s.t.} \quad \mathbf{Ad} = \mathbf{0}$$
$$\left\| \mathbf{Y}^{-1} \mathbf{d} \right\| \leq \beta$$

While linear programming problems do not admit an analytical optimal solution in general, the above optimization problem over an ellipsoid does have an analytical solution interestingly. Under some mathematical conditions, an optimal solution $\mathbf{d}^*$ is:

$$\mathbf{d}^* = -\beta \frac{\mathbf{Y}^2(\mathbf{c} - \mathbf{A}^\top \mathbf{p})}{\left\| \mathbf{Y}(\mathbf{c} - \mathbf{A}^\top \mathbf{p}) \right\|}$$

where $\mathbf{p} = (\mathbf{A}\mathbf{Y}^2\mathbf{A}^\top)^{-1}\mathbf{A}\mathbf{Y}^2\mathbf{c}$. While the analytical solution sounds good, it involves computing inverse matrices, which are very expensive operations in general. Most computational time in the affine scaling algorithm will be used for computing inverse matrices.

With direction $\mathbf{d}^*$, we move to $\mathbf{x} = \mathbf{y} + \mathbf{d}^*$. Then, our next $\mathbf{x}$ remains feasible and interior as well as improve the objective function value. If $\mathbf{c}^\top\mathbf{x} - \mathbf{p}^\top\mathbf{b} < \epsilon$, for some small positive constant ϵ, then we obtain an approximate optimal solution. When $\mathbf{c}^\top\mathbf{x} - \mathbf{p}^\top\mathbf{b}$ is exactly zero, we obtain an exact optimal solution by strong duality. The smaller ϵ is, the closer our solution is to the real optimal solution.

The affine scaling algorithm is presented:

1. Pick some feasible solution $\mathbf{x}^0 > \mathbf{0}$ and set $k = 0$.

2. Given $\mathbf{x}^k > 0$, compute:
$$\mathbf{X}_k = \text{diag}(x_1^k, x_2^k, ..., x_n^k)$$
$$\mathbf{p}^k = (\mathbf{A}\mathbf{X}_k^2\mathbf{A}^\top)^{-1}\mathbf{A}\mathbf{X}_k^2\mathbf{c}$$
$$\mathbf{r}^k = \mathbf{c} - \mathbf{A}^\top\mathbf{p}^k$$

3. If $\mathbf{r}^k \geq \mathbf{0}$ and $(\mathbf{r}^k)^\top\mathbf{x}^k < \epsilon$, then stop; we obtained an (approximately) optimal solution.

4. If $-\mathbf{X_k}^2\mathbf{r}^k \geq \mathbf{0}$, then stop; the problem is unbounded.

5. Update
$$\mathbf{x}^{k+1} = \mathbf{x}^k - \beta\frac{\mathbf{X}_k^2\mathbf{r}^k}{\|\mathbf{X}_k\mathbf{r}^k\|}$$

and repeat.

Suppose we are solving a problem with the following data:

```
c = [-1, -1, 0, 0]
A = [1 2 1 0 ;
     2 1 0 1 ]
b = [3, 3]
```

7.1. The Affine Scaling Algorithm

To use several linear algebra functions, we first need to load the `LinearAlgebra` package, which is one of the standard libraries in Julia.

```
using LinearAlgebra
```

The key implementation part is Step 2 in the algorithm, where we need to perform matrix computations. If `x` is the current $\mathbf{x}^k$, then we can compute $\mathbf{X}_k$, $\mathbf{p}^k$, and $\mathbf{r}^k$ as follows

```
X = Diagonal(x)
p = inv(A*X^2*A')*A*X^2*c
r = c - A'*p
```

In the above code, `Diagonal(x)` constructs a diagonal matrix whose diagonal components are vector `x`. To compute transpose of matrices and vectors, a single quotation mark ' is used. For example, `A'` is the transpose of matrix `A`; alternatively, you can also use `transpose(A)`. To compute the inverse matrix, function `inv()` is used. I believe the code itself is self-explanatory.

Updating `x` can be done as follows:

```
x = x - beta * X^2 * r / norm(X*r)
```

where `norm(X*r)` is used to compute $\|\mathbf{X}_k\mathbf{r}^k\|$.

A complete code for the affine scaling algorithm is shown as a Julia function:

Listing 7.1: The Affine Scaling Algorithm for Linear Programming Problems
code/chap7/affine_scaling.jl

```
using LinearAlgebra

function affine_scaling(c,A,b,x0; beta=0.5, epsilon=1e-9, max_iter=1000)

    # Preparing variables for the trajectories
    x1_traj = []
    x2_traj = []
```

```
    # Initialization
    x = x0

    for i in 1:max_iter
        # Recording the trajectories of x1 and x2
        push!(x1_traj, x[1])
        push!(x2_traj, x[2])

        # Computing
        X = Diagonal(x)
        p = inv(A*X^2*A')*A*X^2*c
        r = c - A'*p

        # Optimality check
        if minimum(r) >= 0  &&  dot(x,r) < epsilon
            break
        end

        # Update
        x = x - beta * X^2 * r / norm(X*r)
    end

    return x1_traj, x2_traj
end
```

Note that `beta`, `epsilon`, and `max_iter` are defined as optional keyword arguments for function `affine_scaling()`. When some function arguments are defined after a semicolon ;, then they become optional keyword arguments. If we call the function as

```
x1_traj, x2_traj = affine_scaling(c, A, b, x0)
```

then `beta=0.5`, `epsilon=1e-9`, and `max_iter=1000` take the default values. If we call the function as

```
x1_traj, x2_traj = affine_scaling(c, A, b, x0, max_iter=50, beta=0.9)
```

then `epsilon=1e-9` take the default values, while `beta` and `max_iter` take the values

7.1. The Affine Scaling Algorithm

supplied. As it is seen in the code, note that the order is not important for optional keyword arguments.

To keep track of the progress, especially the first two components of x, I defined

```
x1_traj = []
x2_traj = []
```

where [] creates an empty array. Then later

```
push!(x1_traj, x[1])
push!(x2_traj, x[2])
```

pushes x[1] and x[2] into the arrays of x1_traj and x2_traj, respectively. After running the algorithm, the last elements of x1_traj and x2_traj will contain the solution obtained. With an initial strictly interior feasible solution x0:

```
x0 = [0.5, 0.03, 2.44, 1.97]
x1_traj, x2_traj = affine_scaling(c, A, b, x0)
```

we obtain an optimal solution:

```
julia> x1_traj[end], x2_traj[end]
(0.9999999820551077, 0.9999999777109649)
```

which is close enough to the exact optimal solution $(1, 1)$.

The trajectories obtained may be plotted using the PyPlot package:

```
using PyPlot
fig = figure()
plot(x1_traj, x2_traj, "o-", label="Affine Scaling")
legend(loc="upper right")
close(fig)
```

The result is shown in Figure 7.1.

Chapter 7. Interior Point Methods

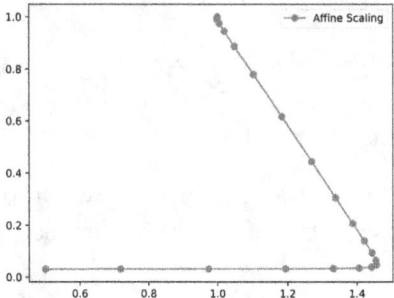

Figure 7.1: The trajectory of the affine scaling algorithm , starting from (0.5, 0.03) and ending at (1.0, 1.0)

7.2 The Primal Path Following Algorithm

In this section, we consider another interior point algorithm, called the primal path following algorithm. We consider a slightly different form of the dual problem. In particular, consider a linear programming problem:

$$\min_{\mathbf{x}} \quad \mathbf{c}^\top \mathbf{x}$$
$$\text{s.t.} \quad \mathbf{A}\mathbf{x} = \mathbf{b}$$
$$\mathbf{x} \geq 0$$

and its dual problem:

$$\max_{\mathbf{p},\mathbf{s}} \quad \mathbf{p}^\top \mathbf{x}$$
$$\text{s.t.} \quad \mathbf{p}^\top \mathbf{A} + \mathbf{s}^\top = \mathbf{c}^\top$$
$$\mathbf{s} \geq 0$$

where **p** and **s** are the dual variables. Note that we have an equality constraint in the dual problem and non-negativity on the slack variable **s**.

In the affine scaling algorithm, we used ellipsoids to remain in the interior of the feasible region. In the primal path following algorithm, we use barrier functions for

157

7.2. The Primal Path Following Algorithm

the non-negativity $\vec{x} \geq \mathbf{0}$. The primal problem with barrier function is:

$$\min_{\mathbf{x}} \ B_\mu(\mathbf{x}) = \mathbf{c}^\top \mathbf{x} - \mu \sum_{j=1}^{n} \log x_j$$

$$\text{s.t.} \quad \mathbf{A}\mathbf{x} = \mathbf{b}$$

With a barrier parameter $\mu > 0$ and $-\log(x_j)$, we impose a big penalty for approaching the boundary $x_j = 0$; therefore, we can expect any solution to the barrier problem is strictly positive.

At any $\mathbf{x} > \mathbf{0}$, using the second-order Taylor expansion, we approximate $B_\mu(\mathbf{x})$ along any direction $\mathbf{d}$:

$$B_\mu(\mathbf{x}+\mathbf{d}) \approx B_\mu(\mathbf{x}) + \sum_{i=1}^{n} \frac{\partial B_\mu(\mathbf{x})}{\partial x_i} d_i + \frac{1}{2} \sum_{i=1}^{n} \sum j=1^{n} \frac{\partial^2 B_\mu(\mathbf{x})}{\partial x_i \partial x_j} d_i d_j$$

$$= B_\mu(\mathbf{x}) + \sum_{i=1}^{n} \left(c_i - \frac{\mu}{x_i} \right) d_i + \frac{1}{2} \sum_{i=1}^{n} \frac{\mu}{(x_i)^2} d_i d_j$$

$$= B_\mu(\mathbf{x}) + (\mathbf{c}^\top - \mu \mathbf{e}^\top \mathbf{X}^{-1}) \mathbf{d} + \frac{1}{2} \mu \mathbf{d}^\top \mathbf{X}^{-2} \mathbf{d}$$

where $\mathbf{X} = \text{diag}(\mathbf{x})$ and $\mathbf{e} = (1, 1, ..., 1)$. Then, we can write the barrier problem as an optimization problem of direction $\mathbf{d}$ as follows:

$$\min_{\mathbf{d}} \ (\mathbf{c}^\top - \mu \mathbf{e}^\top \mathbf{X}^{-1}) \mathbf{d} + \frac{1}{2} \mu \mathbf{d}^\top \mathbf{X}^{-2} \mathbf{d}$$

$$\text{s.t.} \quad \mathbf{A}\mathbf{d} = \mathbf{0}$$

The Lagrangian function for the above problem is constructed:

$$L(\mathbf{d}, \mathbf{p}) = (\mathbf{c}^\top - \mu \mathbf{e}^\top \mathbf{X}^{-1}) \mathbf{d} + \frac{1}{2} \mu \mathbf{d}^\top \mathbf{X}^{-2} \mathbf{d} - \mathbf{p}^\top \mathbf{A} \mathbf{d}$$

where $\mathbf{p}$ is the Lagrangian multiplier. We require

$$\frac{\partial L(\mathbf{d}, \mathbf{p})}{\partial d_j} = 0, \quad \frac{\partial L(\mathbf{d}, \mathbf{p})}{\partial p_i} = 0, \quad \forall i, j$$

for optimality. This leads to the following linear system:

$$\mathbf{c} - \mu \mathbf{X}^{-1} \mathbf{e} + \mu \mathbf{X}^{-2} \mathbf{d} - \mathbf{A}^\top \mathbf{p} = 0$$

or
$$Ad = 0$$

$$\begin{bmatrix} \mu X^{-2} & -A^\top \\ A & 0 \end{bmatrix} \begin{bmatrix} d \\ p \end{bmatrix} = \begin{bmatrix} \mu X^{-1}e - c \\ 0 \end{bmatrix}$$

While closed-form solutions can be derived for the above linear system, we usually solve it using numerical libraries for linear systems based on various factorization techniques as it is.

The primal path following algorithm is presented:

1. Pick some primal feasible solution $x^0 > 0$ and dual feasible solution p and $s^0 > 0$. Select constants $\epsilon > 0$, $\alpha \in (0, 1)$ and $\mu^0 > 0$. Set $k = 0$.

2. If $(s^k)^\top x^k < \epsilon$, then stop; optimal.

3. Let
$$X_k = \text{diag}(x_1^k, x_2^k, ..., x_n^k)$$
$$\mu^{k+1} = \alpha \mu^k$$

4. Solve the linear system:
$$\begin{bmatrix} \mu^{k+1} X_k^{-2} & -A^\top \\ A & 0 \end{bmatrix} \begin{bmatrix} d \\ p \end{bmatrix} = \begin{bmatrix} \mu^{k+1} X_k^{-1} e - c \\ 0 \end{bmatrix}$$

for p and d.

5. Update
$$x^{k+1} = x^k + d$$
$$p^{k+1} = p$$
$$s^{k+1} = c - A^\top p$$

and repeat.

159

7.2. The Primal Path Following Algorithm

While the implementation of the primal path following algorithm is similar to the affine scaling algorithm in most parts, we need to know how to construct and solve the linear system in Step 4. We first define

```
e = ones(length(x),1)
n = length(x)
m = length(b)
```

and

```
X = Diagonal(x)
mu = alpha * mu
```

The left-hand-side and the right-hand-side of the linear system in Step 4 can be constructed as follows:

```
LHS = [ mu*inv(X)^2      -A'          ;
         A               zeros(m,m) ]
RHS = [ mu*inv(X)*e - c ;
         zeros(m,1) ]
```

where `zeros()` is used to construct **0**-matrix and **0**-vector of appropriate size. To solve the linear system, we use the backslash as follows:

```
sol = LHS \ RHS
```

Since `sol` includes solutions for **d** and **p** as a single vector, we split them and update:

```
d = sol[1:n]
p = sol[n+1:end]
x = x + d
s = c - A'*p
```

The complete code is shown:

Listing 7.2: The Primal Path Following Algorithm for Linear Programming Problems
code/chap7/primal_path_following.jl

```julia
using LinearAlgebra

function primal_path_following(c,A,b,x0;
           mu=0.9, alpha=0.9, epsilon=1e-9, max_iter=1000)
  # Preparing variables for the trajectories
  x1_traj = []
  x2_traj = []

  # Initialization
  x = x0
  e = ones(length(x),1)
  n = length(x)
  m = length(b)

  for i=1:max_iter
    # Recording the trajectories of x1 and x2
    push!(x1_traj, x[1])
    push!(x2_traj, x[2])

    # Computing
    X = Diagonal(x)
    mu = alpha * mu

    # Solving the linear system
    LHS = [ mu*inv(X)^2      -A'       ;
              A           zeros(m,m)  ]
    RHS = [ mu*inv(X)*e - c ;
            zeros(m,1) ]
    sol = LHS \ RHS

    # Update
    d = sol[1:n]
    p = sol[n+1:end]
    x = x + d
    s = c - A'*p

    # Optimality check
    if dot(s,x) < epsilon
```

7.3. Remarks

```
        break
    end
end

return x1_traj, x2_traj
end
```

For the same problem, we can compare the trajectories of the affine scaling and the primal path following algorithms.

```
c = [-1, -1, 0, 0]
A = [1 2 1 0 ;
     2 1 0 1 ]
b = [3, 3]

# Initial Starting Solution
x1 = 0.5
x2 = 0.03
# x0 = [0.5, 0.03, 2.44, 1.97]
x0 = [x1, x2, 3-x1-2*x2, 3-2*x1-x2]

x1_traj_a, x2_traj_a = affine_scaling(c, A, b, x0)
x1_traj_p, x2_traj_p = primal_path_following(c, A, b, x0)

using PyPlot
fig = figure()
plot(x1_traj_a, x2_traj_a, "o-", label="Affine Scaling")
plot(x1_traj_p, x2_traj_p, "*-", label="Primal Path Following")
legend(loc="upper right")
savefig("primal_path_following.pdf")
savefig("primal_path_following.png")
close(fig)
```

The result is shown in Figure 7.2.

7.3 Remarks

Similar to the primal path following algorithm, we can also introduce barrier functions for the dual problem as well as for the primal problem. After writing the

Chapter 7. Interior Point Methods

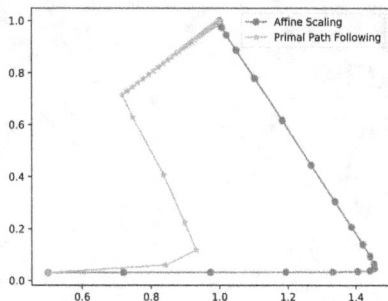

Figure 7.2: The trajectories of the affine scaling and the primal path following algorithms , both starting from (0.5, 0.03) and ending at (1.0, 1.0)

Karush-Kuhn-Tucker (KKT) optimality conditions for the barrier problems, we construct a linear approximation to the KKT conditions, which is solved as a linear system. Then the method is called the primal-dual path following algorithm. Refer to Bertsimas and Tsitsiklis (1997)[2] for more details. Bertsimas and Tsitsiklis (1997) also discuss how we can find an initial interior point.

For nonlinear programming problems, especially for convex optimization problems, interior point methods can be similarly developed. See Nocedal and Writer (2006)[3] for such methods.

Open-source interior point solvers are available in Julia. See Chapter 8 of this book.

---

[2]Bertsimas, D. and Tsitsiklis, J.N., 1997. Introduction to linear optimization. Belmont, MA: Athena Scientific.
[3]Nocedal, J. and Wright, S., 2006. Numerical optimization. Springer Science & Business Media.

7.3. Remarks

8

Nonlinear Optimization Problems

So far, we have learned how we can use the JuMP package for solving linear optimization problem with or without integer variables. In this chapter, I will introduce a few more packages that are useful in solving optimization problems.

8.1 Unconstrained Optimization

We first look at the Optim package[1]. This package implements basic optimization algorithms, mostly for unconstraint nonlinear optimization problems. It contains many standard algorithms that are found in textbooks for nonlinear or numerical optimization such as Bazaraa et al. (2013)[2] and Nocedal and Wright (2006)[3].

8.1.1 Line Search

Algorithms for optimization problems without constraints or with some simple bounds are probably mostly useful in the course of implementing more advanced algorithms. Often algorithms for constrained problems require solving subproblems, which are line search problems in many cases; e.g. in gradient projection algorithms.

---

[1] https://github.com/JuliaNLSolvers/Optim.jl
[2] Bazaraa, M.S., Sherali, H.D. and Shetty, C.M., 2013. Nonlinear programming: theory and algorithms. John Wiley & Sons.
[3] Nocedal, J. and Wright, S., 2006. Numerical optimization. Springer Science & Business Media.

8.1. Unconstrained Optimization

Consider a function $f : \mathbb{R}^2 \mapsto \mathbb{R}$ in the following form:

$$f(\mathbf{x}) = (2x_1 - 3)^4 + (3x_1 - x_2)^2 \tag{8.1}$$

Let $\bar{\mathbf{x}} = (2, 3)$ and a direction $\bar{\mathbf{d}} = (-1, 0)$. Suppose we are solving the following line search problem:

$$\min_{\lambda \in [0,1]} f(\bar{\mathbf{x}} + \lambda \bar{\mathbf{d}})$$

which finds an optimal step size from the current point $\bar{\mathbf{x}}$ in the direction of $\bar{\mathbf{d}}$.

First define:

```
x_bar = [2.0, 3.0]
d_bar = [-1.0, 0.0]
```

We prepare function f in Julia as follows:

```
function f_line(x_bar, d_bar, lambda)
   x_new = x_bar + lambda*d_bar
   return (2x_new[1] - 3)^4 + (3x_new[1] - x_new[2])^2
end
```

Note that `f_line` is indeed a function of `lambda`, while `x_bar` and `d_bar` are regarded as constant parameters, or optional arguments for the purpose of optimization with respect to `lambda`.

To create an ad-hoc function that depends on the current values of `x_bar` and `d_bar` and accepts `lambda` as the only argument, we use

```
lambda -> f_line(x_bar, d_bar, lambda)
```

We now optimize `f_line` in the interval `[0.0, 1.0]` using the Golden Section line search algorithm:

```
using Optim
opt = Optim.optimize( lambda -> f_line(x_bar, d_bar, lambda),
                      0.0, 1.0, GoldenSection())
```

Chapter 8. Nonlinear Optimization Problems

The optimal λ value can be accessed by

```
julia> Optim.minimizer(opt)
0.8489384490968053
```

and the function value at the optimum can be accessed by

```
julia> Optim.minimum(opt)
0.442576659227590126
```

The complete code is shown below:

Listing 8.1: Line Search *code/chap8/linesearch.jl*

```julia
using Optim

function f_line(x_bar, d_bar, lambda)
    x_new = x_bar + lambda*d_bar
    return (2x_new[1] - 3)^4 + (3x_new[1] - x_new[2])^2
end

x_bar = [2.0, 3.0]
d_bar = [-1.0, 0.0]

opt = Optim.optimize( lambda -> f_line(x_bar, d_bar, lambda),
                      0.0, 1.0, GoldenSection())

println("optimal lambda = ", Optim.minimizer(opt))
println("optimal f(x+lambda*d) = ", Optim.minimum(opt))
```

8.1.2 Unconstrained Optimization

Consider the function (8.1) again. This time, we will find a minimum of this function. We prepare function f in Julia as follows:

```julia
function f(x)
    return (2x[1] - 3)^4 + (3x[1] - x[2])^2
end
```

8.1. Unconstrained Optimization

Optimize this function with an initial point [1.0, 3.0]:

```
using Optim
opt = Optim.optimize(f, [1.0, 3.0])
println("optimal x = ", opt.minimizer)
println("optimal f = ", opt.minimum)
```

The result is:

```
optimal x = [1.49581, 4.48742]
optimal f = 4.95862087100503e-9
```

While the default algorithm is Nelder-Mead, other algorithms are also available, such as Simulated Annealing, Broyden–Fletcher–Goldfarb–Shanno (BFGS), Conjugate Gradient, etc:

```
opt = Optim.optimize(f, [1.0, 3.0], NelderMead())
opt = Optim.optimize(f, [1.0, 3.0], SimulatedAnnealing())
opt = Optim.optimize(f, [1.0, 3.0], BFGS())
opt = Optim.optimize(f, [1.0, 3.0], ConjugateGradient())
opt = Optim.optimize(f, [1.0, 3.0], GradientDescent())
```

8.1.3 Box-constrained Optimization

When there are lower and upper bound constraints only, optimization problems are box-constrained. With the function (8.1), consider the following problem:

$$\min \quad f(\mathbf{x})$$
$$\text{s.t.} \quad 2 \leq x_1 \leq 5$$
$$6 \leq x_2 \leq 10$$

We prepare function f in Julia as follows:

```
function f(x)
  return (2x[1] - 3)^4 + (3x[1] - x[2])^2
end
```

We also prepare the lower bound, the upper bound, and an initial solution:

```
lb = [2.0; 6.0]
ub = [5.0; 10.0]
x0 = [3.0; 7.0]
```

Then optimize:

```
Optim.optimize(f, lb, ub, x0, Fminbox(GradientDescent()))
Optim.optimize(f, lb, ub, x0, Fminbox(ConjugateGradient()))
Optim.optimize(f, lb, ub, x0, Fminbox(LBFGS()))
```

For the full list of available algorithms in the `Optim` package, please refer to the official documentation.[4]

8.2 Nonlinear Optimization

A general nonlinear optimization problem, which is potentially nonconvex, can also be modeled by JuMP and solved by `Ipopt`. Consider a nonlinear problem:

$$\min_{x_1, x_2} \quad (x_1 - 3)^3 + (x_2 - 4)^2$$
$$\text{s.t.} \quad (x_1 - 1)^2 + (x_2 + 1)^3 + e^{-x_1} \leq 1$$

These general nonlinear objective function and constraints are handled by macros `@NLobjective` and `@NLconstraint`, respectively. For the above nonlinear optimization problem, we may write:

---
[4] https://github.com/JuliaNLSolvers/Optim.jl

8.3. Other Solvers

```
@NLobjective(m, Min, (x[1]-3)^3 + (x[2]-4)^2)
@NLconstraint(m, (x[1]-1)^2 + (x[2]+1)^3 + exp(-x[1]) <= 1)
```

See the following complete code:

Listing 8.2: NLP with Ipopt *code/chap8/nlp_ipopt.jl*

```
using JuMP, Ipopt
m = Model(Ipopt.Optimizer)

@variable(m, x[1:2])
@NLobjective(m, Min, (x[1]-3)^3 + (x[2]-4)^2)
@NLconstraint(m, (x[1]-1)^2 + (x[2]+1)^3 + exp(-x[1]) <= 1)

JuMP.optimize!(m)

println("** Optimal objective function value = ", JuMP.objective_value(m))
println("** Optimal solution = ", JuMP.value.(x))
```

The result is

```
julia> include("nlp_ipopt.jl")
This is Ipopt version 3.12.10, running with linear solver mumps.
NOTE: Other linear solvers might be more efficient (see Ipopt documentation).

...
...
...

EXIT: Optimal Solution Found.
** Optimal objective function value = 4.409110764366554
** Optimal solution = [0.492399, -0.491889]
```

8.3 Other Solvers

There are many other solvers available in Julia supporting JuMP for various problem structures. See the list of solvers available.[5]

---
[5]http://jump.dev/JuMP.jl/stable/installation/

Chapter 8. Nonlinear Optimization Problems

Some solvers support the AMPL format, also known as `.nl` format. For a complete list of such solvers, visit the website of AMPL.[6] It includes open source solves such as Bonmin and Couenne. Couenne stands for Convex Over and Under ENvelopes for Nonlinear Estimation, and solves nonconvex Mixed Integer Nonlinear Programming (MINLP) problems[7]. Bonmin stands for Basic Open-source Nonlinear Mixed INteger programming and solves general MINLP problems[8].

You can download binary files of Bonmin and Couenne from the download page[9] of AMPL. Save the downloaded files in an appropriate folder and unzip them. For example, the file paths for Bonmin and Couenne in my computer are:

```
/Users/chkwon/ampl/bonmin
/Users/chkwon/ampl/couenne
```

With or without integer variables, Bonmin and Couenne can solve general nonconvex nonlinear optimization problems. As an example, consider a bi-level optimization problem of the form[10]:

$$\begin{aligned}
\min \quad & -x - 3y_1 + 2y_2 \\
\text{s.t.} \quad & -2x + y_1 + 4y_2 + s_1 = 16 \\
& 8x + 3y_1 - 2y_2 + s_2 = 48 \\
& -2x + y_1 - 3y_2 + s_3 = -12 \\
& -y_1 + s_4 = 0 \\
& y_1 + s_5 = 4 \\
& x \geq 0 \\
& -1 + l_1 + 3l_2 + l_3 - l_4 + l_5 = 0 \\
& 4l_2 - 2l_2 - 3l_3 = 0
\end{aligned}$$

---

[6] https://ampl.com/products/solvers/all-solvers-for-ampl/
[7] https://projects.coin-or.org/Couenne
[8] https://projects.coin-or.org/Bonmin
[9] https://ampl.com/products/solvers/open-source/
[10] Ex 9.1.1 from Floudas, C.A., Pardalos, P.M., Adjiman, C., Esposito, W.R., Gümüs, Z.H., Harding, S.T., Klepeis, J.L., Meyer, C.A. and Schweiger, C.A., 2013. Handbook of Test Problems in Local and Global Optimization (Vol. 33). Springer Science & Business Media, http://titan.princeton.edu/TestProblems/

8.3. Other Solvers

$$l_i \geq 0 \quad i = 1, ..., 5$$
$$s_i \geq 0 \quad i = 1, ..., 5$$
$$l_i s_i = 0 \quad i = 1, ..., 5$$

The last constraints $l_i s_i = 0$ are called complementarity conditions, which originate from Karush-Kuhn-Tucker optimality conditions of the original lower-level problem. This class of problems is called a mathematical program with complementarity conditions (MPCC), which is in general a nonconvex, nonlinear optimization problem.

To use these open-source solvers that support the AMPL format, we need to install packages:

```
using Pkg
Pkg.add("AmplNLWriter")
```

The `AmplNLWriter` package provides an interface between JuMP and the solvers. Bonmin and Couenne may be used as follows:

```
using JuMP, AmplNLWriter
m = Model( () -> AmplNLWriter.Optimizer("/Users/chkwon/ampl/bonmin"))
```

or

```
using JuMP, AmplNLWriter
m = Model( () -> AmplNLWriter.Optimizer("/Users/chkwon/ampl/couenne"))
```

First declare variables:

```
@variable(m, x>=0)
@variable(m, y[1:2])
@variable(m, s[1:5]>=0)
@variable(m, l[1:5]>=0)
```

Then, set the objective function:

Chapter 8. Nonlinear Optimization Problems

```
@objective(m, Min, -x -3y[1] + 2y[2])
```

If the objective function were nonlinear, use `@NLobjective(m, Min, ...)`. All linear constraints can be added in the same way as before:

```
@constraint(m, -2x +  y[1] + 4y[2] + s[1] ==  16)
@constraint(m,  8x + 3y[1] - 2y[2] + s[2] ==  48)
@constraint(m, -2x +  y[1] - 3y[2] + s[3] == -12)
@constraint(m,        -y[1]         + s[4] ==   0)
@constraint(m,         y[1]         + s[5] ==   4)
@constraint(m, -1 + l[1] + 3l[2] + l[3] - l[4] + l[5] == 0)
@constraint(m,      4l[2] - 2l[2] - 3l[3]              == 0)
```

Now, to add the nonlinear constraints $l_i s_i = 0$, we use `@NLconstraint`:

```
for i in 1:5
  @NLconstraint(m, l[i] * s[i] == 0)
end
```

After solving the problem by `optimize!(m)`, we obtain the optimal objective function value as -13.0. The complete code is shown below:

Listing 8.3: MPCC example *code/chap8/ex911.jl*

```
using JuMP, AmplNLWriter
m = Model( () -> AmplNLWriter.Optimizer("/Users/chkwon/ampl/bonmin"))
# m = Model( () -> AmplNLWriter.Optimizer("/Users/chkwon/ampl/couenne"))

@variable(m, x>=0)
@variable(m, y[1:2])
@variable(m, s[1:5]>=0)
@variable(m, l[1:5]>=0)

@objective(m, Min, -x -3y[1] + 2y[2])

@constraint(m, -2x +  y[1] + 4y[2] + s[1] ==  16)
@constraint(m,  8x + 3y[1] - 2y[2] + s[2] ==  48)
```

8.3. Other Solvers

```
@constraint(m, -2x  +   y[1] - 3y[2] + s[3] == -12)
@constraint(m,         -y[1]         + s[4] ==   0)
@constraint(m,          y[1]         + s[5] ==   4)
@constraint(m, -1 + l[1] + 3l[2] +  l[3] - l[4] + l[5] == 0)
@constraint(m,      4l[2] - 2l[2] - 3l[3]              == 0)
for i in 1:5
  @NLconstraint(m, l[i] * s[i] == 0)
end

optimize!(m)

println("** Optimal objective function value = ", JuMP.objective_value(m))
println("** Optimal x = ", JuMP.value(x))
println("** Optimal y = ", JuMP.value.(y))
println("** Optimal s = ", JuMP.value.(s))
println("** Optimal l = ", JuMP.value.(l))
```

With `Couenne`, the result is:

```
julia> include("ex911.jl")
Couenne 0.5.6 -- an Open-Source solver for Mixed Integer Nonlinear Optimization
Mailing list: couenne@list.coin-or.org
Instructions: http://www.coin-or.org/Couenne
couenne:
ANALYSIS TEST: NLP0012I
          Num      Status      Obj             It       time      Location
NLP0014I           1         OPT -13            35 0.015721
Couenne: new cutoff value -1.3000000000e+01 (0.022324 seconds)
Loaded instance "/Users/chkwon/.julia/packages/AmplNLWriter/V1gW5/.solverdata/tmpZIiy89.nl"
Constraints:      12
Variables:        13 (0 integer)
Auxiliaries:       6 (0 integer)

Coin0506I Presolve 3 (-6) rows, 4 (-15) columns and 9 (-20) elements
...
...

** Optimal objective function value = -13.0
** Optimal x = 5.0
** Optimal y = [4.0, 2.0]
** Optimal s = [14.0, 0.0, 0.0, 4.0, 0.0]
** Optimal l = [0.0, 0.181818, 0.121212, 0.0, 0.333333]
```

With `Bonmin`, the result is:

```
julia> include("ex911.jl")
Bonmin 1.8.6 using Cbc 2.9.9 and Ipopt 3.12.8
bonmin:
Cbc3007W No integer variables - nothing to do

******************************************************************
```

```
This program contains Ipopt, a library for large-scale nonlinear optimization.
 Ipopt is released as open source code under the Eclipse Public License (EPL).
         For more information visit http://projects.coin-or.org/Ipopt
******************************************************************************

NLP0012I
              Num      Status      Obj             It       time            Location
NLP0014I       1       OPT         -13             24 0.010362
Cbc3007W No integer variables - nothing to do

        "Finished"
** Optimal objective function value = -13.0
** Optimal x = 5.0
** Optimal y = [4.0, 2.0]
** Optimal s = [14.0, 7.52316e-37, 0.0, 4.0, 1.80556e-35]
** Optimal l = [0.0, 0.181818, 0.121212, 0.0, 0.333333]
```

8.4 Mixed Integer Nonlinear Programming

As discussed in the previous section, both Couenne and Bonmin can handle integer variables. Let all variables in the MPCC example from the previous section to integer variables, and solve again.

```
@variable(m, x>=0, Int)
@variable(m, y[1:2], Int)
@variable(m, s[1:5]>=0, Int)
@variable(m, l[1:5]>=0, Int)
```

We will obtain the same objective function value, -13.0, with slightly different optimal solutions:

```
Bonmin 1.8.6 using Cbc 2.9.9 and Ipopt 3.12.8
bonmin:

******************************************************************************
This program contains Ipopt, a library for large-scale nonlinear optimization.
 Ipopt is released as open source code under the Eclipse Public License (EPL).
         For more information visit http://projects.coin-or.org/Ipopt
******************************************************************************

NLP0012I
              Num      Status      Obj             It       time            Location
NLP0014I       1       OPT         -13             24 0.010299
NLP0012I
              Num      Status      Obj             It       time            Location
NLP0014I       1       OPT         -13             8 0.003689
NLP0014I       2       OPT         -13             0 0
NLP0012I
              Num      Status      Obj             It       time            Location
NLP0014I       1       OPT         -13             0 0
Cbc0012I Integer solution of -13 found by DiveMIPFractional after 0 iterations and 0 nodes (0.00 seconds)
Cbc0001I Search completed - best objective -13, took 0 iterations and 0 nodes (0.00 seconds)
```

8.4. Mixed Integer Nonlinear Programming

```
Cbc0035I Maximum depth 0, 0 variables fixed on reduced cost

        "Finished"
** Optimal objective function value = -13.0
** Optimal x = 5.0
** Optimal y = [4.0, 2.0]
** Optimal s = [14.0, 0.0, 0.0, 4.0, 0.0]
** Optimal l = [0.0, 0.0, 0.0, 0.0, 1.0]
```

We see that `Bonmin` uses `Cbc` and `Ipopt` internally.

9 Monte Carlo Methods

Monte Carlo methods are a group of computational methods based on random number generation. In most cases, what we do in Monte Carlo methods is in the following form:

1. Generate random numbers;

2. Solve some problems or compute some values deterministically given the random numbers generated; and

3. Estimate some quantities, usually by computing weighted averages of the results from deterministic computation.

This chapter will introduce how we can generate random numbers in Julia and illustrate the method applied in revenue management and estimating the number of paths in a network.

9.1 Probability Distributions

We use the Distributions[1] package. Our objective is to generate 100 random numbers from Normal distribution with mean $\mu = 50$ and standard deviation $\sigma = 7$. First install, if you haven't done yet, the Distributions package and import it:

---

[1] https://github.com/JuliaStats/Distributions.jl

9.1. Probability Distributions

```
using Pkg
Pkg.add("Distributions")
using Distributions
```

Then create a Normal distribution:

```
d = Normal(50,7)
```

Draw 100 random variable from the distribution d:

```
x = rand(d, 100)
```

Other standard distributions are also available. For example, a discrete distribution, Bernoulli distribution with $p = 0.25$ can be used as follows:

```
d = Bernoulli(0.25)
x = rand(d, 100)
```

Available distributions are listed in the following document: Univariate Distributions[2]

Suppose now that we have a dataset, and we want to fit a distribution to this dataset and estimate the parameters. For example, we have

```
data = [2, 3, 2, 1, 4, 2, 1, 1, 2, 1]
```

and want to fit with a Binomial distribution. We use the `fit()` function:

```
julia> fit(Geometric, data)
Geometric{Float64}(p=0.3448275862068966)
```

To create a multi-variate Normal distribution, we can use the following code:

---

[2]https://juliastats.org/Distributions.jl/latest/univariate/

```
mean = [10.0; 20.1]
covariance_matrix = [2.0 1.2; 1.2 3.5]
d = MvNormal(mean, covariance_matrix)
x = rand(d, 5)
```

where we generate 5 samples. Note that `covariance_matrix` is a symmetric matrix. The result is

```
julia> x = rand(d, 5)
2x5 Array{Float64,2}:
  8.11396   9.0993   8.18729  12.6574   9.0898
 18.5511   21.8543  17.6941   19.8019  17.6679
```

We can access the third sample in the third column as follows:

```
julia> x[:,3]
2-element Array{Float64,1}:
  8.18729
 17.6941
```

9.2 Randomized Linear Program

In this section, I introduce an application of Monte Carlo methods in revenue management. This section is based on the excellent textbook on the topic by Talluri and van Ryzin (2006)[3] and the original paper by Talluri and van Ryzin (1999)[4].

You don't need to fully understand the context of revenue management is to see how Julia can be used to perform tasks in Monte Carlo simulations. I'll just give a short description of a revenue management problem for those who are interested in.

Consider an airline service company that provides several different service products $j = 1, ..., n$. For each product j, the company needs to determine how many

---

[3] Talluri, K.T. and Van Ryzin, G.J., 2006. The Theory and Practice of Revenue Management (Vol. 68). Springer Science & Business Media.

[4] Talluri, K. and Van Ryzin, G., 1999. A Randomized Linear Programming Method for Computing Network Bid Prices. Transportation Science, 33(2), pp.207-216.

9.2. Randomized Linear Program

tickets to sell, which is denoted by y_j. For each product j, the demand is a random variable D_j. There is a resource constraint

$$\mathbf{Ay} \le \mathbf{x}$$

where $A_{ij} = 1$ if resource i is used by product j and $\mathbf{x}$ is a vector of available resources. Let us denote the price for each product by p_j. Given $\mathbf{x}$ and the random demand $\mathbf{D}$, the company considers the following assignment problem:

$$H(\mathbf{x}, \mathbf{D}) = \max \quad \mathbf{p}^\top \mathbf{y}$$
$$\text{s.t.} \quad \mathbf{Ay} \le \mathbf{x}$$
$$0 \le \mathbf{y} \le \mathbf{D}$$

Note that function $H(\mathbf{x}, \mathbf{D})$, which is the optimal objective function value given $\mathbf{x}, \mathbf{D}$, is a random varialbe, since $\mathbf{D}$ is a random variable. Let $\pi(\mathbf{x}, \mathbf{D})$ denote the dual variable for the resource constraint, which is also a random variable.

In this capacity control problem, what we are interested is to compute:

$$\pi = \nabla_x \mathbb{E}[H(\mathbf{x}, \mathbf{D})]$$

Under some mathematical conditions, the gradient and the expection operators can be interchanged:

$$\pi = \nabla_x \mathbb{E}[H(\mathbf{x}, \mathbf{D})] = \mathbb{E}[\nabla_x H(\mathbf{x}, \mathbf{D})] = \mathbb{E}[\pi(\mathbf{x}, \mathbf{D})]$$

To estimate the expected value of dual variables, we use the Monte Carlo simulation. We have the following tasks:

1. To generate N samples of $\mathbf{D}$, namely

$$\mathbf{D}^{(1)}, \mathbf{D}^{(2)}, ..., \mathbf{D}^{(N)};$$

2. To solve the optimization problem for each $\mathbf{D}^{(k)}$, $k = 1, ..., N$, and call the dual variable for the resource constraint $\pi^{(k)}$; and

3. To obtain an estimate:

$$\pi \approx \frac{1}{N} \sum_{k=1}^{N} \pi^{(k)}.$$

Chapter 9. Monte Carlo Methods

If you cannot understand the context of this problem, that's fine; my explanation was short and insufficient. If you are interested in this application, I again refer to the excellent textbook by Talluri and van Ryzin (2006) mentioned earlier. Anyway, we will see how we can complete the above tasks 1, 2, and 3 in Julia.

Consider a simple airline network: Tampa (TPA) → New York (JFK) → Frankfrut (FRA). There are two resources:

Resource	Origin → Destination
1	TPA → JFK
2	JFK → FRA

We have the following six products:

Product	Description	Revenue	Demand (μ, σ^2)
1	TPA to JFK full fare	$150	$(30, 5^2)$
2	TPA to JFK discount	$100	$(60, 7^2)$
3	JFK to FRA full fare	$120	$(20, 2^2)$
4	JFK to FRA discount	$80	$(80, 4^2)$
5	TPA to JFK to FRA full fare	$250	$(30, 3^2)$
6	TPA to JFK to FRA discount	$170	$(40, 9^2)$

Normal distributions $N(\mu, \sigma^2)$ are assumed for demand. Suppose the remaining capacity for each resource is given by:

Resource	Remaining Capacity
1	120
2	230

To implement the Monte Carlo simulation to estimate π in Julia, let's consider the following basic structure:

1. Random number generation

```
samples = Array{Float64}(undef, no_products, N)
```

We'll *somehow* generate N samples of demand for each product and store it in an array named `samples`.

9.2. Randomized Linear Program

2. Solving deterministic problems

```
pi_samples = Array{Float64}(undef, no_resources, N)
for k in 1:N
  pi_samples[:,k] = DLP(x, samples[:,k])
end
```

We will create a function DLP that solves a determinisitc LP problem and returns the optimal dual variable values, which will be stored in an array named pi_samples.

3. Estimating π

```
pi_estimate = sum(pi_samples, dims=2) / N
```

This is a simple way of computing the average. I will explain what the expression sum(pi_samples, dims=2) means later.

First, we import packages that we need:

```
using JuMP, GLPK, Distributions
```

The above table is converted to arrays in Julia: for products

```
# Product Data
no_products = 6
products = 1:no_products
p = [150; 100; 120; 80; 250; 170]
mu = [30; 60; 20; 80; 30; 40]
sigma = [5; 7; 2; 4; 3; 9]
```

for resources

```
no_resources = 2
resources = 1:no_resources
x = [110; 230]
```

The product-resource incident matrix A may be written as:

```
A = [ 1 1 0 0 1 1 ;
      0 0 1 1 1 1 ]
```

We design a Julia function that accepts **x** and **D** as input and returns the vector of dual variables π as output:

```julia
function DLP(x, D)
  m = Model(GLPK.Optimizer)
  @variable(m, y[products] >= 0)
  @objective(m, Max, sum( p[j]*y[j] for j in products) )

  # Resource Constraint
  @constraint(m, rsc_const[i in 1:no_resources],
          sum( A[i,j]*y[i] for j in products) <= x[i] )

  # Upper Bound
  @constraint(m, bounds[j in 1:no_products], y[j] <= D[j] )

  JuMP.optimize!(m)
  pi_val = JuMP.shadow_price.(rsc_const)
  return pi_val
end
```

The above function DLP should be self-explaining. We should however note that variables `products`, `p`, `no_resources`, `no_products`, and `A` must be accessible within the scope block of function DLP; that is, these variables should have been defined within the parent scope block. Please keep this in mind, when you review my complete code later.

Now, we are ready to run the Monte Carlo simulation.

1. Generate random number for the demand as follows:

```julia
N = 100
samples = Array{Float64}(undef, no_products, N)
for j in products
   samples[j,:] = rand( Normal(mu[j], sigma[j]), N)
end
```

9.2. Randomized Linear Program

where `samples[j,:]` is the j-th row of `samples` and `samples[j,k]` is the k-th sample of demand for product j.

2. For each sample, solve the optimization problem and obtain the dual variable value:

```
pi_samples = Array{Float64}(undef, no_resources, N)
for k in 1:N
    pi_samples[:,k] = DLP(x, samples[:,k])
end
```

where `pi_samples[:,k]` is the k-th column of `pi_samples` and the element `pi_samples[i,k]` is the dual variable for resource i with k-th demand sample.

3. Compute an estimate of π:

```
pi_estimate = sum(pi_samples, dims=2) / N
```

where `sum(pi_samples, dims=22)` computes summation over the second dimension of the `pi_samples` array. Just to see what exactly it is, observe:

```
julia> pi_samples
2x100 Array{Float64,2}:
 0.0  37.5  37.5   0.0   0.0  ...  37.5  37.5  37.5  37.5  37.5
 0.0   0.0   0.0  25.0  25.0       0.0  25.0   0.0  25.0   0.0

julia> sum(pi_samples, dims=1)
1x100 Array{Float64,2}:
 0.0  37.5  37.5  25.0  25.0  ...  37.5  62.5  37.5  62.5  37.5

julia> sum(pi_samples, dims=2)
2x1 Array{Float64,2}:
 2550.0
 1425.0
```

The complete code is shown below:

Listing 9.1: Monte Carlo Simulation for Randomized LP *code/chap9/rlp.jl*

```julia
# Importing packages
using JuMP, GLPK, Distributions

# Product Data
no_products = 6
products = 1:no_products
p = [150; 100; 120; 80; 250; 170]
mu = [30; 60; 20; 80; 30; 40]
sigma = [5; 7; 2; 4; 3; 9]

# Resource Data
no_resources = 2
resources = 1:no_resources
x = [110; 230]

# A_ij = 1, if resource i is used by product j
A = [ 1 1 0 0 1 1 ;
      0 0 1 1 1 1 ]


# Solving the deterministic LP problem
function DLP(x, D)
  m = Model(GLPK.Optimizer)
  @variable(m, y[products] >= 0)
  @objective(m, Max, sum( p[j]*y[j] for j in products) )

  # Resource Constraint
  @constraint(m, rsc_const[i in 1:no_resources],
          sum( A[i,j]*y[j] for j in products) <= x[i]  )

  # Upper Bound
  @constraint(m, bounds[j in 1:no_products], y[j] <= D[j] )

  JuMP.optimize!(m)
  pi_val = JuMP.shadow_price.(rsc_const)
  return pi_val
end

# Generating N samples
N = 100
samples = Array{Float64}(undef, no_products, N)
```

9.3. Estimating the Number of Simple Paths

```
for j in products
  samples[j,:] = rand( Normal(mu[j], sigma[j]), N)
end

# Obtain the dual variable for each sample
pi_samples = Array{Float64}(undef, no_resources, N)
for k in 1:N
  pi_samples[:,k] = DLP(x, samples[:,k])
end

# Compute the average
pi_estimate = sum(pi_samples, dims=2) / N
println("** pi estimate = ", pi_estimate')
```

The result is

```
julia> include("rlp.jl")
** pi estimate = [26.625 13.5]
```

We obtained $\pi_1 = 26.625$ and $\pi_2 = 13.5$, which are called bid prices. This results suggest that the airline company rejects booking requests for the first resource—flight from TPA to JFK—with the revenue less than \$26.625, and the second resource—flight from JFK to FRA—less than \$13.5. I wish these prices were real.

Since this is a Monte Carlo simulation, the results would change everytime you run the code. With very large N, the result would be more accurate.

9.3 Estimating the Number of Simple Paths

A path is *simple* if each node in the path is visited only once. In a network, finding the exact number of all available simple paths can take very long time. For small-scale networks, we can simply enumerate all paths.

Consider the network in Figure 9.1. There are 5 nodes and 8 undirected links. If we enumerate all simple paths from node 1 to node 5, we obtain the following 9 simple paths:

$$\{1,2,3,4,5\}$$

Chapter 9. Monte Carlo Methods

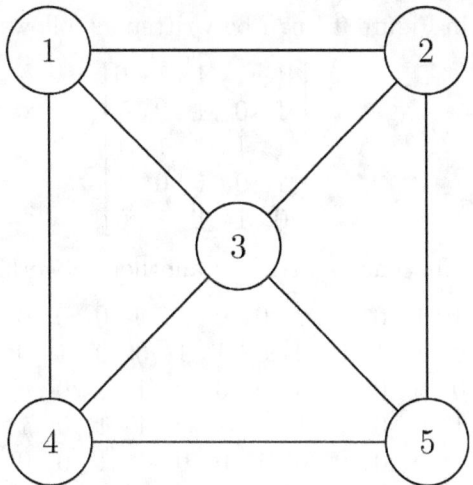

Figure 9.1: A simple network with 5 nodes

$$\{1, 2, 3, 5\}$$
$$\{1, 2, 5\}$$
$$\{1, 3, 2, 5\}$$
$$\{1, 3, 4, 5\}$$
$$\{1, 3, 5\}$$
$$\{1, 4, 3, 2, 5\}$$
$$\{1, 4, 3, 5\}$$
$$\{1, 4, 5\}$$

Let $\mathcal{P}$ denote the above set of all simple paths and $|\mathcal{P}|$ the number of all simple paths. In the above example, $|\mathcal{P}| = 9$.

One may express the network structure by an node-arc adjacency matrix $\mathbf{A}$, where $A_{ij} = 1$ if there is a link from node i to node j. For example, the adjacency

9.3. Estimating the Number of Simple Paths

matrix of the network in Figure 9.1 can be written as follows:

$$\mathbf{A} = \begin{bmatrix} 0 & 1 & 1 & 1 & 0 \\ 1 & 0 & 1 & 0 & 1 \\ 1 & 1 & 1 & 1 & 1 \\ 1 & 0 & 1 & 0 & 1 \\ 0 & 1 & 1 & 1 & 0 \end{bmatrix} \tag{9.1}$$

Then, consider the adjacency matrix for another network:

$$\mathbf{A} = \begin{bmatrix} 0 & 1 & 1 & 0 & 0 & 1 & 0 & 0 & 1 & 0 & 0 & 0 & 0 & 0 & 0 & 0 \\ 0 & 0 & 1 & 1 & 0 & 0 & 0 & 1 & 1 & 0 & 0 & 1 & 0 & 1 & 1 & 1 \\ 1 & 0 & 0 & 0 & 0 & 0 & 0 & 0 & 1 & 1 & 0 & 0 & 0 & 0 & 1 \\ 1 & 1 & 0 & 0 & 1 & 1 & 1 & 1 & 0 & 1 & 1 & 0 & 1 & 0 & 0 & 1 \\ 0 & 1 & 0 & 0 & 0 & 0 & 1 & 0 & 0 & 1 & 1 & 0 & 0 & 0 & 1 & 0 \\ 1 & 0 & 0 & 0 & 0 & 0 & 0 & 1 & 0 & 0 & 1 & 0 & 0 & 0 & 0 \\ 0 & 0 & 0 & 1 & 0 & 1 & 0 & 0 & 1 & 0 & 0 & 0 & 1 & 0 & 1 \\ 1 & 0 & 0 & 0 & 0 & 0 & 0 & 1 & 1 & 1 & 0 & 0 & 1 & 0 & 0 \\ 1 & 1 & 0 & 0 & 1 & 0 & 0 & 0 & 0 & 1 & 0 & 0 & 1 & 0 & 0 \\ 1 & 1 & 0 & 0 & 1 & 1 & 0 & 0 & 1 & 0 & 0 & 0 & 1 & 0 & 1 & 0 \\ 0 & 0 & 1 & 0 & 0 & 0 & 0 & 0 & 1 & 0 & 0 & 0 & 0 & 0 & 1 \\ 0 & 0 & 1 & 1 & 1 & 1 & 0 & 1 & 0 & 0 & 1 & 0 & 1 & 1 & 0 & 1 \\ 1 & 1 & 0 & 0 & 1 & 1 & 0 & 0 & 0 & 0 & 0 & 0 & 1 & 0 & 0 \\ 0 & 0 & 1 & 0 & 1 & 1 & 1 & 0 & 1 & 0 & 0 & 1 & 0 & 0 & 1 & 0 \\ 1 & 0 & 0 & 0 & 0 & 0 & 0 & 0 & 0 & 0 & 1 & 1 & 0 & 0 & 1 \\ 1 & 1 & 0 & 1 & 0 & 0 & 0 & 1 & 0 & 0 & 1 & 0 & 0 & 0 & 0 & 0 \end{bmatrix} \tag{9.2}$$

There are 16 nodes and 89 directed links. How many simple paths between node 1 to node 16? Well, there are 138,481 simple paths; hence $|\mathcal{P}| = 138,481$. The number of simple paths increase exponentially. If the size of the network gets bigger and bigger, counting all of simple paths one by one will take very long time. If we need a good estimate for $|\mathcal{P}|$, we can use a Monte Carlo method, as suggested by Roberts and Kroese (2007)[5].

Like in other Monte Carlo methods, we will also generate many samples—in this case *sample paths*—for estimating $|\mathcal{P}|$. To illustrate this process, let's use the example network in Figure 9.1.

---

[5]Roberts, B., & Kroese, D. P. (2007). Estimating the Number of s-t Paths in a Graph. Journal of Graph Algorithms and Applications, 11(1), 195–214.

1. First we begin with the origin node 1. There are three nodes that are connected to it: nodes 2, 3, and 4. Randomly choose a node with some probability. Let's use a Uniform distribution. Say, node 4 is chosen with probability 1/3. Our current path is $\{1, 4\}$.

2. Node 4 is connected to nodes 1, 3, and 5. Since node 1 is already visited, we choose between nodes 3 and 5, randomly. Say, node 3 is chosen with probability 1/2. The current path is $\{1, 4, 3\}$.

3. Node 3 is connected to *unvisited* nodes 2 and 5. Again randomly choose between nodes 2 and 5. Say, node 5 is chosen with probability 1/2. The current path is $\{1, 4, 3, 5\}$.

4. Since we arrived at the destination node 5, we stop.

By the above process, one sample path is generated. Let us denote this first sample by $p^{(1)}$. The probability at which this sample path is generated is $g(p^{(1)}) = \frac{1}{3} \times \frac{1}{2} \times \frac{1}{2} = \frac{1}{12}$. We can repeat this process N times to generate N sample paths. In each sample generation, we stop if either the destination node is reached or there is no more unvisited node connected to the current node. We obtain:

$$p^{(1)}, \quad p^{(2)}, \quad \ldots, \quad p^{(N)}$$
$$g(p^{(1)}), \ g(p^{(2)}), \ \ldots, \ g(p^{(N)})$$

Among the above N samples, there may be some samples that didn't reach the destination node. In the network in Figure 9.1, we will always end up with the destination node, but in general, we don't.

To estimate the number of all simple paths based on the above N samples, we compute the following quantity:

$$\widehat{|\mathcal{P}|} = \frac{1}{N} \sum_{i=1}^{N} \frac{\mathbb{I}[p^{(i)} \text{ reached the destination}]}{g(p^{(i)})}$$

where $\mathbb{I}[\cdot]$ is a function whose value is 1 if the condition specified in the brackets is true and 0 otherwise. The expression $\mathbb{I}[p^{(i)}$ reached the destination] may also be written as $\mathbb{I}[p^{(i)} \in \mathcal{P}]$. With sufficiently large N, we have $\widehat{|\mathcal{P}|} \approx |\mathcal{P}|$.

To implement this Monte Carlo method in Julia, we first build a function with the following input and output:

9.3. Estimating the Number of Simple Paths

- Input: the adjacency matrix, the origin node, and the destination node
- Output: whether the path generated reached the destination or not $\mathbb{I}[p^{(i)} \in \mathcal{P}]$ and the sample's probability $g(p^{(i)})$.

That is, we have a function like

```
function generate_sample_path(adj_mtx, origin, destination)
    # after some calculations

    return I, g
end
```

where `adj_mtx` stores the adjacency matrix in the form of (9.1) and (9.2).

Let's fill in the function `generate_sample_path()`. I'll first make a copy of `adj_mtx`:

```
adj_copy = copy(adj_mtx)
```

I want to make changes in the values of `adj_mtx` in the process, but still want to preserve the original value of `adj_mtx`. Note that if you directly change the value of `adj_mtx` even inside the function `generate_sample_path()`, you still change the *original* `adj_mtx`. So I make a copy.

Prepare some initial values

```
path = [origin]
g = 1
current = origin
```

where `path` is an array that I will keep adding node numbers to the end, `g` is the probability value to which I will keep multiplying a probability, and `current` will track at which node we are currently located.

Since the `origin` node is already visited, I'll just disconnect the node from all other nodes in the *copied* adjacency matrix `adj_copy` as follows:

```
adj_copy[:, origin] .= 0
```

which essentially makes the first column of `adj_copy` zero, so that moving from any other node to the origin node impossible.

Now I'll repeat some computations until we reach the destination node, i.e. `current==destination`. Therefore we prepare a `while` loop:

```
while current != destination
  # Do some updates on path, g, and current
end
```

In the above loop, we will search nodes that are connected to the current node and randomly choose one of them. Note that we don't need to worry if each node is visited or unvisited, since all visited nodes will be disconnected in the `adj_copy` before we search them.

Inside the loop, we will store all nodes that are connected to `current` in an array `V`:

```
# Inside the loop
V = findall(adj_copy[current,:] .== 1)
```

where `findall(x .== 1)` returns the indices of unitary elements of array x. In this case, we are essentially finding node i such that `adj_copy[current,i] == 1`. Just to make it sure that we understand what `findall` does, see the following example:

```
julia> x = [1; 1; 0; 0; 0; 1]
julia> findall(x .== 1)
3-element Array{Int64,1}:
 1
 2
 6
```

If there is no element in `V`, there is no way to proceed; we stop the loop. The sample path we just generated didn't reach the destination.

9.3. Estimating the Number of Simple Paths

```
# Inside the loop
if length(V)==0
   break
end
```

Now we choose a node from **V** randomly using a Uniform distribution and add it to **path**:

```
# Inside the loop
next = rand(V)
path = [path; next]
```

Before we proceed to the next iteratin of the loop, we update three variables:

```
# Inside the loop
current = next
adj_copy[:, next] = 0
g = g / length(V)
```

The first line updates the current node, the second line disconnects the node we just randomly chose from all other nodes, and the third line update the probability **g**.

The entire **while** loop looks:

```
while current != destination
   # Find all nodes connected to current
   V = findall(adj_copy[current,:] .==1 )
   if length(V)==0
      break
   end

   # Choose a node randomly and add to path
   next = rand(V)
   path = [path; next]

   # Update variables for the next iteration
   current = next
   adj_copy[:, next] = 0
   g = g / length(V)
end
```

The above `while` loop will end either because it reached the destination node or because there was no more connected nodes to visit. After the loop, we set the value of I either 1 or 0, and return it with the value of g as follows:

```
I = 0
if path[end]==destination
  I = 1
end

return I, g
```

The complete code is shown below:

Listing 9.2: Generating a Sample Path *code/chap9/generate_sample_path.jl*

```
function generate_sample_path(adj_mtx, origin, destination)
  # Make a copy of adj_mtx
  adj_copy = copy(adj_mtx)

  # Initializing variables
  path = [origin]
  g = 1
  current = origin

  # Disconnect origin from all other nodes
  adj_copy[:, origin] .= 0

  while current != destination
    # Find all nodes connected to current
    V = findall(adj_copy[current,:] .==1 )
    if length(V)==0
      break
    end

    # Choose a node randomly and add to path
    next = rand(V)
    path = [path; next]

    # Update variables for the next iteration
    current = next
```

9.3. Estimating the Number of Simple Paths

```
        adj_copy[:, next] .= 0
        g = g / length(V)
    end

    I = 0
    if path[end]==destination
        I = 1
    end

    return I, g
end
```

Calling this function many times, we can generate as many sample paths as we want. Let's test it for the adjacency matrix shown in (9.1):

```
adj_mtx = [
0 1 1 1 0 ;
1 0 1 0 1 ;
1 1 1 1 1 ;
1 0 1 0 1 ;
0 1 1 1 0
]
origin = 1
destination = 5

N = 100
estimate_samples = Array{Float64}(undef, N)
for i in 1:N
    I, g = generate_sample_path(adj_mtx, origin, destination)
    estimate_samples[i] = I / g
end
est = sum(estimate_samples) / N
println("estimate = ", est)
```

In the above code, we generate 100 samples. It's weird to generate 100 paths when there are only 9 paths, but this is a test. The result is:

```
estimate = 8.85
```

which will change every time we run the code.

For the adjacency matrix shown in (9.2), if we generate 1,000 samples, the result looks like:

```
estimate = 135253.784
```

This *simple* and *naive* Monte Carlo method for estimating $|\mathcal{P}|$ tends to generate shorter paths; hence we need a more advanced sampling technique to generate longer paths more frequently than in the naive sampling, which is called *importance sampling*. Interested readers should read the paper by Roberts and Kroese (2007)[6].

In fact, I have already implemented the algorithm suggested by Roberts and Kroese (2007) and written a package for it. Please check the `PathDistribution` package[7].

---

[6] Roberts, B., & Kroese, D. P. (2007). Estimating the Number of s-t Paths in a Graph. Journal of Graph Algorithms and Applications, 11(1), 195–214.

[7] https://github.com/chkwon/PathDistribution.jl

9.3. Estimating the Number of Simple Paths

10
Lagrangian Relaxation

Computational methods based on Lagrangian relaxation have been popularly used for solving many optimization problems, especially for mixed integer linear programming problems and combinatorial optimization problems. The key idea behind the algorithm is typically that we relax hard constraints to make the relaxed problems relatively easier to solve. While we solve these easy problems multiple times, we update the values of Lagrangian multipliers adequately.

In this chapter, I'll briefly introduce how methods based on Lagrangian relaxation work and present a few applications with Julia codes.

10.1 Introduction

An excellent introduction the Lagrangian relaxation method is provided by Fisher (2004)[1]. This section briefly summarizes the method.

Consider an integer programming problem:

$$\begin{aligned} Z^* = \min \quad & \mathbf{c}^\top \mathbf{x} \\ \text{s.t.} \quad & \mathbf{A}\mathbf{x} = \mathbf{b} \\ & \mathbf{D}\mathbf{x} \leq \mathbf{e} \end{aligned}$$

---

[1] Fisher, M.L., 2004. The Lagrangian relaxation method for solving integer programming problems. Management Science, 50(12_supplement), pp.1861-1871.

10.1. Introduction

$$\mathbf{x} \geq \mathbf{0}, \text{ Integer.}$$

Let λ denote the dual variable for the first equality constraint. By relaxing the first constraint, we formulate the Lagrangian problem

$$\begin{aligned} Z_D(\lambda) = \min \quad & \mathbf{c}^\top \mathbf{x} + \lambda^\top (\mathbf{A}\mathbf{x} - \mathbf{b}) \\ \text{s.t.} \quad & \mathbf{D}\mathbf{x} \leq \mathbf{e} \\ & \mathbf{x} \geq \mathbf{0}, \text{ Integer.} \end{aligned}$$

The Lagrangian problem is relatively easier to solve than the original problem.

A brief structure of the Lagrangian relaxation method can be described as follows:

- **Step 0:** Guess an initial value of λ^0.

- **Step 1:** Given the current value of λ^k, solve the Lagrangian problem and obtain $Z_D(\lambda^k)$. Let the solution to the Lagrangian problem $\mathbf{x}_D^k$, which is likely infeasible to the original problem.

- **Step 2:** Given the current values of λ^k and $\mathbf{x}_D^k$, obtain a feasible solution $\mathbf{x}^k$, *somehow*.

- **Step 3:** Update λ^k to λ^{k+1}, *somehow*, and repeat.

There are two *somehow*'s in the above procedure. I'll explain each somehow one by one.

10.1.1 Lower and Upper Bounds

By solving the original problem, we aim to compute the optimal objective function value Z^*, which is unknown at this moment. Our goal is to bound this unknown optimal objective function value both from below and above. That is, we want to obtain lower bound Z_{LB} and upper bound Z_{UB} such that

$$Z_{\text{LB}} \leq Z^* \leq Z_{\text{UB}}$$

We become confident that we obtained an optimal solution when $Z_{\text{UB}} = Z_{\text{LB}}$. The difference between these two bounds is called the optimality gap, and typically calculated by the following percentage:

$$\text{Optimality Gap} = \frac{Z_{\text{UB}} - Z_{\text{LB}}}{Z_{\text{UB}}} \times 100\%$$

In the Lagrangian relaxation method described earlier, it is well known that

$$Z_D(\lambda) \leq Z^*$$

for all $\lambda \geq \mathbf{0}$. Therefore, the relaxed Lagrangian problem provides lower-bounds. We want the lower bounds are close to the optimal Z^*. To do so, we need to make a good choice of λ. This is done by updating λ^k wisely. I'll explain this in the next section.

We can consider another relaxation, called LP relaxation in the following form:

$$\begin{aligned} Z_{\text{LP}} = \min \quad & \mathbf{c}^\top \mathbf{x} \\ \text{s.t.} \quad & \mathbf{A}\mathbf{x} = \mathbf{b} \\ & \mathbf{D}\mathbf{x} \leq \mathbf{e} \\ & \mathbf{x} \geq \mathbf{0} \end{aligned}$$

where we just eliminate the integrality condition. The LP relaxation also provides an lower bound:

$$Z_{\text{LP}} \leq Z^*$$

which is usually not very good; i.e. not close enough to Z^*.

To find an upper bound, we just need to find a feasible solution to the original problem, since

$$Z^* \leq \mathbf{c}^\top \mathbf{x}$$

for any feasible $\mathbf{x}$ by the definition of Z^*. The real question is how to find a *good* upper bound—as close as possible to the optimal solution. A possible approach is to use the solution to the Lagrangian problem. When we solve the Lagrangian problem given the multiplier λ^k, we obtain a solution $\mathbf{x}_D^k$, which is infeasible; otherwise we find an optimal solution. One may fix some variables at the value of $\mathbf{x}_D^k$ and then modify other variables to create a feasible solution $\mathbf{x}^k$, or apply some heuristic algorithms to the original problem starting from the Lagrangian solution $\mathbf{x}_D^k$ to obtain a feasible solution $\mathbf{x}^k$. This requires developments of methods that are specific to the application on the table.

10.1. Introduction

10.1.2 Subgradient Optimization

When updating λ^k, we basically want to maximize the lower bound. That is we solve

$$\max_{\lambda} Z_D(\lambda)$$

To solve this problem, we use a subgradient optimization method. The updating scheme is

$$\lambda^{k+1} = \lambda^k + t_k(\mathbf{A}\mathbf{x}_D^k - \mathbf{b})$$

where $t_k > 0$ is a step size. If you had relaxed inequality constraints, you should enforce λ^{k+1} nonnegative or nonpositive, depending on the direction of the inequality. In this example, however, since we relaxed equality constraints, the sign of λ is unrestricted. The most popular choice for the step size is

$$t_k = \frac{\theta_k(Z_{\text{UB}} - Z_D(\lambda^k))}{||\mathbf{A}\mathbf{x}_D^k - b||^2}$$

where Z_{UB} is the best-known—smallest—upper bound and $\theta_k \in (0, 2]$ is a scalar. Some use a fixed λ at all iterations, while some use a varying λ in each iteration depending on the progress of the algorithm.

10.1.3 Summary

In developing a Lagrangian relaxation method, we need to make three decisions:

1. Which constraints to be relaxed,

2. How to obtain a feasible solution and obtain an upper bound, and

3. How to update the Lagrangian multiplier.

The first decision on selecting constraints to be relaxed impacts the overall computational difficulties and the solution quality. Depending on the choice of constraints, the Lagrangian problem may be very easy to solve or may remain to be challenging. All of the above three decisions should be specific to the particular application of interest.

10.2 The p-Median Problem

In this section, we apply the Lagrangian relaxation method to solve a facility location problem, called the p-median problem and how we can write a Julia code. A compact introduction to the subject is provided by Daskin and Maass (2015)[2].

Suppose there are n customers who are geographically apart. We want to determine the locations of p number of facilities to serve these customers. Each facility is assumed to have infinite service capacity. That is, we can simply assign each customer to exactly one facility. Our objective is to minimize the weighted sum of transportation cost to serve all customers.

We first define some sets:

- $\mathcal{I} = \{1, ..., m\}$: the set of candidate locations for facilities
- $\mathcal{J} = \{1, ..., n\}$: the set of customer locations

and some parameters:

- p: the number of facilities to introduce
- d_j: the size of demand at customer $j \in \mathcal{J}$
- c_{ij}: the transportation cost from facility location $i \in \mathcal{I}$ to the customer location $j \in \mathcal{J}$

We now introduce variables:

- x_{ij}: the fraction of demand from customer $j \in \mathcal{J}$ is served by facility location $i \in \mathcal{I}$
- y_i: location variable with $y_i = 1$ if a facility is introduced at candidate location $i \in \mathcal{I}$ and $y_i = 0$ otherwise.

The p-median problem can be formulated as follows:

$$Z^* = \min \sum_{i \in \mathcal{I}} \sum_{j \in \mathcal{J}} d_j c_{ij} x_{ij}$$

---

[2] Daskin, M.S. and Maass, K.L., 2015. The p-median problem. In Location Science (pp. 21-45). Springer International Publishing. http://doi.org/10.1007/978-3-319-13111-5_2

10.2. The p-Median Problem

$$\text{s.t.} \quad \sum_{i \in \mathcal{I}} x_{ij} = 1 \quad \forall j \in \mathcal{J}$$

$$\sum_{i \in \mathcal{I}} y_i = p$$

$$x_{ij} \leq y_i \quad \forall i \in \mathcal{I}, j \in \mathcal{J}$$

$$y_i \in \{0, 1\} \quad \forall i \in \mathcal{I}$$

$$x_{ij} \geq 0 \quad \forall i \in \mathcal{I}, j \in \mathcal{J}$$

While this may be solved well by CPLEX and Gurobi for smaller networks, it takes too much time for larger problems; in general the p-median problem is NP-hard. Let's first solve a small instance optimally.

10.2.1 Reading the Data File

We need to prepare the two sets $\mathcal{I}$ and $\mathcal{J}$ and three parameters p, **d** and **c**. As a small example, we consider 10 customers and 7 candidate locations. As explained in Section 3.9, we will use .csv files to prepare data and read them into Julia.

The demand data file `demand.csv` should look like:

```
demand
10
6
20
32
15
28
3
19
8
13
```

The transportation cost **c** in the table form may be:

Chapter 10. Lagrangian Relaxation

	C1	C2	C3	C4	C5	C6	C7	C8	C9	C10
L1	10	7	11	12	32	15	20	26	4	41
L2	13	17	31	37	21	5	13	15	14	12
L3	4	13	14	22	8	31	26	11	12	23
L4	21	21	13	18	9	27	11	16	26	32
L5	32	18	11	14	11	11	16	32	34	8
L6	15	9	13	12	14	15	32	8	12	9
L7	28	32	15	2	17	12	9	6	11	6

In the above table 'C1' means customer 1 and 'L1' means candidate location 1. The corresponding `cost.csv` file should look like:

```
,C1,C2,C3,C4,C5,C6,C7,C8,C9,C10
L1,10,7,11,12,32,15,20,26,4,41
L2,13,17,31,37,21,5,13,15,14,12
L3,4,13,14,22,8,31,26,11,12,23
L4,21,21,13,18,9,27,11,16,26,32
L5,32,18,11,14,11,11,16,32,34,8
L6,15,9,13,12,14,15,32,8,12,9
L7,28,32,15,2,17,12,9,6,11,6
```

Read the `.csv` files and put them in arrays:

```
using DelimitedFiles
d, header = readdlm("demand.csv", ',', header=true)
data = readdlm("cost.csv", ',')
cc = data[2:end, 2:end]
c = convert(Array{Float64,2}, cc)
```

Note in the intermediate form `cc` is of `Array{Any,2}` type:

```
julia> typeof(cc)
Array{Any,2}
```

It is `Any`, because `data` has text strings and numbers mixed together. To use it with other mathematical functions that deal with numerical data, we convert it to `Array{Float64,2}`, which is a two-dimensional array of `Float64` type:

10.2. The p-Median Problem

```
julia> c = convert(Array{Float64,2}, cc)
7x10 Array{Float64,2}:
 10.0   7.0  11.0  12.0  32.0  15.0  20.0  26.0   4.0  41.0
 13.0  17.0  31.0  37.0  21.0   5.0  13.0  15.0  14.0  12.0
  4.0  13.0  14.0  22.0   8.0  31.0  26.0  11.0  12.0  23.0
 21.0  21.0  13.0  18.0   9.0  27.0  11.0  16.0  26.0  32.0
 32.0  18.0  11.0  14.0  11.0  11.0  16.0  32.0  34.0   8.0
 15.0   9.0  13.0  12.0  14.0  15.0  32.0   8.0  12.0   9.0
 28.0  32.0  15.0   2.0  17.0  12.0   9.0   6.0  11.0   6.0
```

We now have the `d` vector and the `c` matrix ready.

Just as a small error-check, we see if the length of `d` and the number of columns in `c` match:

```
@assert length(d) == size(c,2)
```

The `@assert` macro is useful for this purpose. It tests the given statement and produces an error if the given statement is `false`. If the given statement is `true`, it just does nothing and proceeds to the next line of the code.

Then we create ranges for the two sets $\mathcal{I}$ and $\mathcal{J}$:

```
locations = 1:size(c,1)  # the set, I
customers = 1:length(d)  # the set, J
```

10.2.2 Solving the p-Median Problem Optimally

We will create a Julia function that solves the p-median problem optimally for a given p and returns the optimal solution. We will use the `JuMP` and `Gurobi` packages. The function is named `optimal(p)` and looks like the following:

```
function optimal(p)
  m = Model(GLPK.Optimizer)

  @variable(m, x[i in locations, j in customers] >= 0)
  @variable(m, y[i in locations], Bin)
```

```
@objective(m, Min, sum( d[j]*c[i,j]*x[i,j]
                for i in locations, j in customers) )

@constraint(m, [j in customers], sum( x[i,j] for i in locations) == 1)
@constraint(m, sum( y[i] for i in locations) == p)
@constraint(m, [i in locations, j in customers], x[i,j] <= y[i] )

JuMP.optimize!(m)

Z_opt = JuMP.objective_value(m)
x_opt = JuMP.value.(x)
y_opt = JuMP.value.(y)

return Z_opt, x_opt, y_opt
end
```

I believe the above code is easy to understand; if not you may want to read previous chapters again, especially Chapter 2.

10.2.3 Lagrangian Relaxation

To apply the Lagrangian relaxation method, we relax the first constraint that enforces the total assignment for each demand is 100% and obtain the following Lagrangian problem:

$$Z_D(\lambda) = \min \quad \sum_{i \in \mathcal{I}} \sum_{j \in \mathcal{J}} d_j c_{ij} x_{ij} + \sum_{j \in \mathcal{J}} \lambda_j \left(1 - \sum_{i \in \mathcal{I}} x_{ij}\right)$$

$$= \min \quad \sum_{i \in \mathcal{I}} \sum_{j \in \mathcal{J}} (d_j c_{ij} - \lambda_j) x_{ij} + \sum_{j \in \mathcal{J}} \lambda_j$$

$$\text{s.t.} \quad \sum_{i \in \mathcal{I}} y_i = p$$

$$x_{ij} \leq y_i \quad \forall i \in \mathcal{I}, j \in \mathcal{J}$$

$$y_i \in \{0, 1\} \quad \forall i \in \mathcal{I}$$

$$x_{ij} \geq 0 \quad \forall i \in \mathcal{I}, j \in \mathcal{J}$$

10.2. The p-Median Problem

10.2.4 Finding Lower Bounds

The above Lagrangian problem has a special structure that makes it easy to solve. Due to the relaxation, we don't necessarily need to assign all demands. The following simple procedure finds an optimal solution to the Lagrangian problem for any given λ:

1. **Step 1:** For all $i \in \mathcal{I}$, compute $v_i = \sum_{j \in \mathcal{J}} \min\{0, d_j c_{ij} - \lambda_j\}$.

2. **Step 2:** Sort candidate locations by the value of v_i, and select the p most negative v_i values. (Ties can be broken arbitrarily.)

3. **Step 3:** Set $y_i = 1$ for the chosen candidate locations.

4. **Step 4:** Set $x_{ij} = 1$ if $y_i = 1$ and $d_j c_{ij} - \lambda_j < 0$.

Note that solving the Lagrangian problem and finding lower bounds do not even require solving an optimization problem; we just need to sort some values.

Step 1 is straightforward. We first declare y as an array of type `Float64` and the same size as `locations`. Then we compute as indicated in Step 1 and assign values.

```
v = Array{Float64}(undef, size(locations))
for i in locations
  v[i] = 0
  for j in customers
    v[i] = v[i] + min(0, d[j]*c[i,j] - lambda[j] )
  end
end
```

One may rewrite the above code with a fewer lines with some vectorization, but the above is just easy to understand. It is anyway faster with for-loops without vectorization in Julia. In MATLAB, vectorization is always much faster than for-loops.

In Step 2, to select the p most negative v_i values, we will use a sorting function, namely `sortperm()`, which returns the ordered array index, rather than the sorted values themselves.

```
idx = sortperm(v)
```

For example, v has the following values:

```
julia> v
7-element Array{Float64,1}:
  -8.0
  -1.0
   0.0
  -7.0
   0.0
   0.0
 -13.0
```

If we want to choose the $p = 3$ most negative values, we have to choose v[7], v[1], and v[4]. The result of sortperm(v) is:

```
julia> sortperm(v)
7-element Array{Int64,1}:
 7
 1
 4
 2
 3
 5
 6
```

We are interested in the first three values of sortperm(v), which are 7, 1, and 4. There are many different functions are prepared for our convenience. See the official documentation on functions related to sorting[3].

In Step 3, we first prepare a zero vector and let $y_i = 1$ for those chosen locations i in Step2:

---

[3]https://docs.julialang.org/en/v1/base/sort/

10.2. The p-Median Problem

```
y = zeros(Int, size(locations))
y[idx[1:p]] .= 1
```

In the above example `idx[1:p]` has the following values:

```
julia> idx[1:p]
3-element Array{Int64,1}:
 7
 1
 4
```

The resulting y array looks like:

```
julia> y
7-element Array{Int64,1}:
 1
 0
 0
 1
 0
 0
 1
```

which is exactly what we wanted.

In Step 4, based on the value stored in the y array, we determine the value of x.

```
x = zeros(Int, length(locations), length(customers))
for i in locations, j in customers
  if y[i]==1 && d[j]*c[i,j]-lambda[j]<0
    x[i,j] = 1
  end
end
```

In the above code, the symbol `&&` means 'and'. When both of the two statements are true, the line with `x[i,j] = 1` is executed.

We also need to compute the value of $Z_D(\lambda)$:

```
Z_D = 0.0
for j in customers
  Z_D = Z_D + lambda[j]
  for i in locations
    Z_D = Z_D + d[j]*c[i,j]*x[i,j] - lambda[j]*x[i,j]
  end
end
```

which is essentially identical to

$$Z_D(\lambda) = \min \; \sum_{i \in \mathcal{I}} \sum_{j \in \mathcal{J}} d_j c_{ij} x_{ij} + \sum_{j \in \mathcal{J}} \lambda_j \left(1 - \sum_{i \in \mathcal{I}} x_{ij} \right)$$

The complete code for finding an lower bound is prepared as a function:

```
function lower_bound(lambda)
  # Step 1: Computing v
  v = Array{Float64}(undef, size(locations))
  for i in locations
    v[i] = 0
    for j in customers
      v[i] = v[i] + min(0, d[j]*c[i,j] - lambda[j] )
    end
  end

  # Step 2: Sorting v from the most negative to zero
  idx = sortperm(v)

  # Step 3: Determine y
  y = zeros(Int, size(locations))
  y[idx[1:p]] .= 1

  # Step 4: Determine x
  x = zeros(Int, length(locations), length(customers))
  for i in locations, j in customers
    if y[i]==1 && d[j]*c[i,j]-lambda[j]<0
      x[i,j] = 1
    end
  end
```

10.2. The p-Median Problem

```
# Computing the Z_D(lambda^k)
Z_D = 0.0
for j in customers
    Z_D = Z_D + lambda[j]
    for i in locations
        Z_D = Z_D + d[j]*c[i,j]*x[i,j] - lambda[j]*x[i,j]
    end
end

return Z_D, x, y
end
```

which accepts `lambda` and `p` as inputs and returns `Z_D`, `x` and `y`.

10.2.5 Finding Upper Bounds

By solving the Lagrangian problem, we have the location variable **y** and the assignment variable **x** on our hands. While the location variable **y** satisfies all the original constraints, the assignment variable **x** likely violate the original constraint that is relaxed ($\sum_{i \in \mathcal{I}} x_{ij} = 1$ for all $j \in \mathcal{J}$). To find a feasible solution and obtain an upper bound, we fix **y** at the solution of the Lagrangian problem, then assign each demand node to the nearest facility. This again simple procedure will provide us with an upper bound at each iteration.

The Julia function for finding upper bounds may be written as follows:

```
function upper_bound(y)
    # Computing x, given y
    x = zeros(Int, length(locations), length(customers))
    for j in customers
        idx = argmin( c[:,j] + (1 .- y) .* maximum(c) )
        x[idx,j] = 1
    end

    # Computing Z
    Z = 0.0
    for i in locations
        for j in customers
            Z = Z + d[j]*c[i,j]*x[i,j]
        end
```

```
        end

    return Z, x
end
```

The first part computes a feasible solution **x**, and the second parts computes the objective function value at **x**; hence it provides an upper bound. While the second part is straightforward, the first part needs some explanation. In the above code, `c[:,j]` is j-th column of the c matrix. We want to pick the smallest value among the values in `c[:,j]` with `y[i]=1`. There will be many different ways of doing this. My strategy is to add a big number to the elements of `c[:,j]` with `y[i]=0`; that is, `c[:,j] + (1-y)*maximum(c)`. When j=1, the original data `c[:,1]` looks like:

```
julia> c[:,1]
7-element Array{Float64,1}:
 10.0
 13.0
  4.0
 21.0
 32.0
 15.0
 28.0
```

If `y=[1; 0; 0; 1; 0; 0; 1]`, we see

```
julia> c[:,1] + (1 .- y) .* maximum(c)
7-element Array{Float64,1}:
 10.0
 54.0
 45.0
 21.0
 73.0
 56.0
 28.0
```

Then we can simply choose the smallest value from the above result, to select the nearest *open* facility location from the customer. The function `argmin(vector)` finds the index of the minimum value in `vector`; therefore we obtain

10.2. The p-Median Problem

```
julia> idx = argmin( c[:,j] + (1 .- y) .* maximum(c) )
1
```

which indicates that the nearest open facility to customer 1 is in location 1.

10.2.6 Updating the Lagrangian Multiplier

To update the Lagrangian multiplier λ^k at each iteration k, we use the following subgradient optimization method:

$$\lambda_j^{k+1} = \lambda_j^k + t_k \left(1 - \sum_{i \in \mathcal{I}} x_{Dij}^k \right)$$

The step size t_k is determined as follows:

$$t_k = \frac{\theta_k (Z_{\text{UB}} - Z_D(\lambda^k))}{\sum_{j \in \mathcal{J}} \left(1 - \sum_{i \in \mathcal{I}} x_{Dij}^k \right)^2}$$

where Z_{UB} is the best-known—smallest—upper bound, x_{Dij}^k is the the element at (i, j) of the solution $\mathbf{x}_D^k$ to the Lagrangian problem at iteration k, and $\theta_k \in (0, 2]$ is a scalar.

To run the iterations of the Lagrangian relaxation algorithm by updating the multiplier as described above, we need some preparation in Julia. We first determine the maximum number of iterations to allow:

```
MAX_ITER = 10000
```

We also prepare two array objects to track what lower and upper bounds are obtained at each iteration:

```
UB = Array{Float64}(undef, 0)
LB = Array{Float64}(undef, 0)
```

These two array objects contain nothing at this moment. The current best lower and upper bounds are recorded in the following two scalar variables:

```
Z_UB = Inf
Z_LB = -Inf
```

Initial values for the upper and lower bounds are set to ∞ and $-\infty$, respectively. We also record the current best feasible solution (from the current best upper bound):

```
x_best = zeros(length(locations), length(customers))
y_best = zeros(length(locations))
```

Finally we prepare an initial guess on the Lagrangian multiplier λ:

```
lambda = zeros(size(customers))
```

where we set $\lambda_j = 0$ for all $j \in \mathcal{J}$.

To run the iterations of the Lagrangian relaxation method, we will use for-loop:

```
for k=1:MAX_ITER
    ...
    if opt_gap < 0.000001
        break
    end
end
```

The loop will terminate once the optimality gap `opt_gap` is less than a small number, 0.000001.

Inside the loop, we do the following. First we obtain lower and upper bounds, given the current value of `lambda`:

```
Z_D, x_D, y = lower_bound(lambda, p)
Z, x = upper_bound(y)
```

We then update the best-known upper and lower bounds, and the best-known feasible solutions:

10.2. The p-Median Problem

```
# Updating the upper bound
if Z < Z_UB
    Z_UB = Z
    x_best = x
    y_best = y
end

# Updating the lower bound
if Z_D > Z_LB
    Z_LB = Z_D
end
```

Just to observe the progress of the algorithm, we add the upper and lower bounds obtained from the current iteration to the track record array objects UB and LB:

```
push!(UB, Z)
push!(LB, Z_D)
```

Note that the function push!() adds the new value to the *end* of the array; hence it will change the array variable itself—that's why the function has ! in its name.

We now compute the step size and update the multiplier. In this example, we just set $\theta_k = 1.0$ for the simplicity:

```
theta = 1.0
```

We use an array object residual to compute a vector whose j-th element is $1 - \sum_{i \in \mathcal{I}} x^k_{Dij}$:

```
residual = 1 .- transpose(sum(x_D, dims=1))
```

where sum(x_D, 1) returns a row vector with $\sum_{i \in \mathcal{I}} x^k_{Dij}$ as its elements. Note that when we do addition or subtraction between a scalar and a vector, Julia automatically assumes that the scalar is a vector of all equal elements. That is 1 in the above code is converted to ones(length(customers),1). The next step is to the multiplier:

```
t = theta * (Z_UB - Z_D) / sum(residual.^2)
lambda = lambda + t * residual
```

where the .^2 means element-wise square; that is, sum(residual.^2) is equivalent to $\sum_{j \in \mathcal{J}} \left(1 - \sum_{i \in \mathcal{I}} x_{Dij}^k\right)^2$

The Lagrangian relaxation method should now look like:

```
function lagrangian_relaxation(p)
    # The maximum number of iterations allowed
    MAX_ITER = 10000

    # To track the upper and lower bounds
    UB = Array{Float64}(undef, 0)
    LB = Array{Float64}(undef, 0)

    # The best-known upper and lower bounds
    Z_UB = Inf
    Z_LB = -Inf

    # The best-known feasible solutions
    x_best = zeros(length(locations), length(customers))
    y_best = zeros(length(locations))

    # Initial multiplier
    lambda = zeros(size(customers))

    for k=1:MAX_ITER
        # Obtaining the lower and upper bounds
        Z_D, x_D, y = lower_bound(lambda)
        Z, x = upper_bound(y)

        # Updating the upper bound
        if Z < Z_UB
            Z_UB = Z
            x_best = x
            y_best = y
        end

        # Updating the lower bound
        if Z_D > Z_LB
            Z_LB = Z_D
```

10.2. The p-Median Problem

```
        end

        # Adding the bounds from the current iteration to the record
        push!(UB, Z)
        push!(LB, Z_D)

        # Determining the step size and updating the multiplier
        theta = 1.0
        residual = 1 .- transpose(sum(x_D, dims=1))
        t = theta * (Z_UB - Z_D) / sum(residual.^2)
        lambda = lambda + t * residual

        # Computing the optimality gap
        opt_gap = (Z_UB-Z_LB) / Z_UB
        if opt_gap < 0.000001
            break
        end
    end

    return Z_UB, x_best, y_best, UB, LB
end
```

When implementing the Lagrangian relaxation methods, one should pay attention to the variables regarding the bounds: one should not be confused among Z, Z_D, Z_LB, and Z_UB. Also be aware that if the original problem is minimization or maximization and the relaxed constraint is equality or inequality.

Let's first solve the problem optimally:

```
Z_opt, x_opt, y_opt = optimal(p)
```

which yields:

```
julia> Z_opt
1029.0

julia> y_opt
1-dimensional JuMPArray{Float64,1,...} with index sets:
    Dimension 1, 1:7
And data, a 7-element Array{Float64,1}:
```

Chapter 10. Lagrangian Relaxation

```
0.0
1.0
1.0
0.0
0.0
0.0
1.0
```

Let's check if the Lagrangian relaxation method found a good solution:

```
julia> Z_UB, x_best, y_best, UB, LB = lagrangian_relaxation(p)

julia> Z_UB
1029.0

julia> y_best
7-element Array{Int64,1}:
 0
 1
 1
 0
 0
 0
 1
```

Comparing the above two values with `Z_opt` and `y_opt`, we see that the Lagrangian relaxation found the same optimal solution. Let's also see how it progressed:

```
julia> [LB UB]
34x2 Array{Float64,2}:
    0.0     1382.0
  464.2     1073.0
  880.933   1181.0
  934.547   1029.0
  924.036   1072.0
  939.354   1073.0
  964.13    1029.0
  977.495   1072.0
  984.033   1073.0
  996.843   1029.0
```

217

10.2. The p-Median Problem

```
1001.92    1382.0
 998.96    1072.0
1004.45    1073.0
1009.86    1072.0
1016.51    1382.0
1021.49    1073.0
1023.58    1029.0
1025.88    1072.0
1026.57    1382.0
1026.57    1029.0
1027.02    1073.0
1027.38    1072.0
1027.97    1029.0
1028.24    1072.0
1028.62    1029.0
1028.67    1073.0
1028.89    1072.0
1028.86    1029.0
1028.93    1072.0
1028.97    1029.0
1028.98    1072.0
1028.99    1029.0
1029.0     1072.0
1029.0     1029.0
```

Note that the optimal solution was found as early as in the fourth iteration. However, it went over many more iterations to confirm that it is indeed an optimal solution. The lower bound kept increasing and reached at the same bound in the 34-th iteration.

Using the `PyPlot` package as described in Section 3.10.1, we can provide a plot as in Figure 10.1.

The complete code is presented below:

Listing 10.1: The Lagrangian Relaxation Method for the p-Median Problem
code/chap10/p-median.jl

```julia
using JuMP, GLPK, DelimitedFiles, PyPlot

# Solving the p-median problem by Lagrangian Relaxation
p = 3
```

Chapter 10. Lagrangian Relaxation

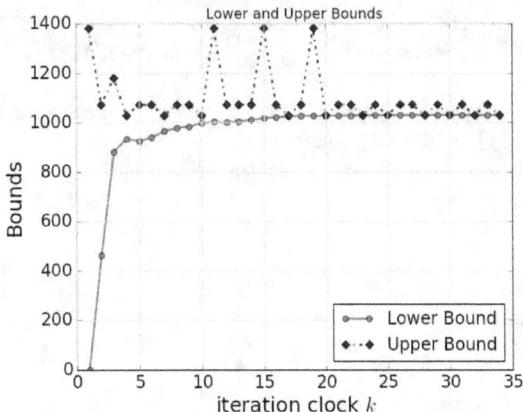

Figure 10.1: Iterations of the Lagrangian Relaxation Method

```
# Reading demand data
d, header = readdlm("demand.csv", ',', header=true)

# Reading transportation cost data
data = readdlm("cost.csv", ',')
cc = data[2:end, 2:end]
c = convert(Array{Float64,2}, cc)

# the length of 'd' and the number of columns in 'c' must match
@assert length(d) == size(c,2)

locations = 1:size(c,1) # the set, I
customers = 1:length(d) # the set, J

# making these data global so that any function can access data
global locations, customers, c


function optimal(p)
  m = Model(GLPK.Optimizer)

  @variable(m, x[i in locations, j in customers] >= 0)
  @variable(m, y[i in locations], Bin)
```

219

10.2. The p-Median Problem

```julia
    @objective(m, Min, sum( d[j]*c[i,j]*x[i,j]
                    for i in locations, j in customers) )

    @constraint(m, [j in customers], sum( x[i,j] for i in locations) == 1)
    @constraint(m, sum( y[i] for i in locations) == p)
    @constraint(m, [i in locations, j in customers], x[i,j] <= y[i] )

    JuMP.optimize!(m)

    Z_opt = JuMP.objective_value(m)
    x_opt = JuMP.value.(x)
    y_opt = JuMP.value.(y)

    return Z_opt, x_opt, y_opt
end


function lower_bound(lambda)
    # Step 1: Computing v
    v = Array{Float64}(undef, size(locations))
    for i in locations
        v[i] = 0
        for j in customers
            v[i] = v[i] + min(0, d[j]*c[i,j] - lambda[j] )
        end
    end

    # Step 2: Sorting v from the most negative to zero
    idx = sortperm(v)

    # Step 3: Determine y
    y = zeros(Int, size(locations))
    y[idx[1:p]] .= 1

    # Step 4: Determine x
    x = zeros(Int, length(locations), length(customers))
    for i in locations, j in customers
        if y[i]==1 && d[j]*c[i,j]-lambda[j]<0
            x[i,j] = 1
        end
    end
```

```
    # Computing the Z_D(lambda^k)
    Z_D = 0.0
    for j in customers
      Z_D = Z_D + lambda[j]
      for i in locations
        Z_D = Z_D + d[j]*c[i,j]*x[i,j] - lambda[j]*x[i,j]
      end
    end

    return Z_D, x, y
end


function upper_bound(y)
    # Computing x, given y
    x = zeros(Int, length(locations), length(customers))
    for j in customers
      idx = argmin( c[:,j] + (1 .- y) .* maximum(c) )
      x[idx,j] = 1
    end

    # Computing Z
    Z = 0.0
    for i in locations
      for j in customers
        Z = Z + d[j]*c[i,j]*x[i,j]
      end
    end
    return Z, x
end

function lagrangian_relaxation(p)
    # The maximum number of iterations allowed
    MAX_ITER = 10000

    # To track the upper and lower bounds
    UB = Array{Float64}(undef, 0)
    LB = Array{Float64}(undef, 0)

    # The best-known upper and lower bounds
    Z_UB = Inf
```

10.2. The p-Median Problem

```julia
Z_LB = -Inf

# The best-known feasible solutions
x_best = zeros(length(locations), length(customers))
y_best = zeros(length(locations))

# Initial multiplier
lambda = zeros(size(customers))

for k=1:MAX_ITER
  # Obtaining the lower and upper bounds
  Z_D, x_D, y = lower_bound(lambda)
  Z, x = upper_bound(y)

  # Updating the upper bound
  if Z < Z_UB
    Z_UB = Z
    x_best = x
    y_best = y
  end

  # Updating the lower bound
  if Z_D > Z_LB
    Z_LB = Z_D
  end

  # Adding the bounds from the current iteration to the record
  push!(UB, Z)
  push!(LB, Z_D)

  # Determining the step size and updating the multiplier
  theta = 1.0
  residual = 1 .- transpose(sum(x_D, dims=1))
  t = theta * (Z_UB - Z_D) / sum(residual.^2)
  lambda = lambda + t * residual

  # Computing the optimality gap
  opt_gap = (Z_UB-Z_LB) / Z_UB
  if opt_gap < 0.000001
    break
  end
end
```

```
        return Z_UB, x_best, y_best, UB, LB
end


# Finding the exact optimal solution
Z_opt, x_opt, y_opt = optimal(p)
# Finding a solution by Lagrangian relaxation
Z_UB, x_best, y_best, UB, LB = lagrangian_relaxation(p)



iter = 1:length(LB)
fig = figure()

# Plotting two datasets
plot(iter, LB, color="red", linewidth=2.0, linestyle="-",
    marker="o", label="Lower Bound")
plot(iter, UB, color="blue", linewidth=2.0, linestyle="-.",
    marker="D", label="Upper Bound")

# Labeling axes
xlabel(L"iteration clock $k$", fontsize="xx-large")
ylabel("Bounds", fontsize="xx-large")

# Putting the legend and determining the location
legend(loc="lower right", fontsize="x-large")

# Add grid lines
grid(color="#DDDDDD", linestyle="-", linewidth=1.0)
tick_params(axis="both", which="major", labelsize="x-large")

# Title
title("Lower and Upper Bounds")
savefig("iterations.png")
savefig("iterations.pdf")
close(fig)
```

10.2. The p-Median Problem

11

Complementarity Problems

This chapter covers modeling and solution approaches for various equilibrium problems arising in various problems, mainly in economics, operations research, regional science, and transportation engineering. We will focus on (linear, nonlinear, or mixed) complementarity problems and introduce tools available in Julia.

A complementarity problem has the following structure:

$$\mathbf{F}(\mathbf{z}) \geq \mathbf{0}, \qquad \mathbf{z} \geq \mathbf{0}, \qquad \mathbf{F}_j(\mathbf{z})z_j = 0 \; \forall j = 1, ..., n$$

where $z \in \mathbb{R}^n$ and $F : \mathbb{R}^n \mapsto \mathbb{R}^n$. When $F(\cdot)$ is a linear function, then the problem is called a Linear Complementarity Problem (LCP); when it is a nonlinear function, the problem is called a Nonlinear Complementarity Problem (NCP). When the lower bounds are not zero, or there is an upper bound, the problem is called a Mixed Complementarity Problem (MCP).

11.1 Linear Complementarity Problems (LCP)

An LCP is typically formulated as follows:

$$\mathbf{w} = \mathbf{Mz} + \mathbf{q} \geq \mathbf{0}, \qquad \mathbf{z} \geq \mathbf{0}, \qquad w_j z_j = 0 \; \forall j = 1, ..., n$$

where M is a $n \times n$ matrix and q, z, and w are n-dimensional vectors. The complementarity condition $w_j z_j = 0$ can also be written in the vector form:

$$\mathbf{w}^\top \mathbf{z} = 0.$$

11.1. Linear Complementarity Problems (LCP)

The following notation is also popularly used:

$$0 \leq \mathbf{z} \perp \mathbf{Mz} + \mathbf{q} \geq \mathbf{0}.$$

We consider a simple production equilibrium problem as an illustrative example for LCP. Suppose that Player 1's production planning problem is defined as follows:

$$\max_{q_1 \geq 0} f_1(q_1; q_2) = p(q) \cdot q_1 - c_1 q_1$$

where q_1 is the production quantity of Player 1, q_2 is the production quantity of Player 2, $p(q)$ is the market sales price function with $q = q_1 + q_2$, and c_1 is the unit production cost of Player 1. Note that Player 1's problem assumes that the production quantity of Player 2, q_2, is given.

Similarly, Player 2's problem is:

$$\max_{q_2 \geq 0} f_2(q_2; q_1) = p(q) \cdot q_2 - c_2 q_2$$

where the market sales price $p(q)$ is shared with Player 1.

The market price is a function of the total production from the two players. In particular, we assume

$$p(q) = a - b(q_1 + q_2)$$

for some positive constants a and b. With this price function, both objective functions f_1 and f_2 are concave quadratic functions.

From the perspective of Player 1, the optimal production level q_1 can be determined as follows. Suppose q_2 is given. If there exists q_1 such that

$$\frac{\partial f_1(q_1; q_2)}{\partial q_1} = 0, \qquad q_1 \geq 0$$

then such q_1 is an optimal solution. If not, the following must hold at $q_1 = 0$:

$$\frac{\partial f_1(0; q_2)}{\partial q_1} \leq 0.$$

The partial derivative is

$$\frac{\partial f_1(q_1; q_2)}{\partial q_1} = a - c_1 - 2bq_1 - bq_2.$$

With the above two conditions together, we can write:

$$-(a - c_1 - 2bq_1 - bq_2)q_1 = 0, \quad q_1 \geq 0, \quad -(a - c_1 - 2bq_1 - bq_2) \geq 0.$$

Similarly, for Player 2:

$$-(a - c_2 - 2bq_2 - bq_1)q_2 = 0, \quad q_2 \geq 0, \quad -(a - c_2 - 2bq_2 - bq_1) \geq 0.$$

The above two conditions form an LCP.

To model the above LCP in Julia, we use the Complementarity package[1]. This package utilizes the JuMP package and connects with external solvers available in Julia such as PATHSolver.jl and NLsolve.jl. The PATHSolver package uses the PATH Solver[2]. Note that the PATH Solver requires a license to solve large problems with more than 300 variables and 2,000 non-zeros. For a temporary license valid for a year, visit the website of PATH Solver: http://pages.cs.wisc.edu/~ferris/path.html. The NLsolve package provides native Julia solvers for a system of nonlinear equations, of which complementarity problems are special cases.

First, add the Complementarity package:

```
using Pkg; Pkg.add("Complementarity")
```

which will also add PATHSolver.jl and NLsolve.jl.

To solve the LCP from the production equilibrium problem, we prepare data:

```
a = 10; b = 1; c1 = 3; c2 = 4
```

We begin with

```
using Complementarity, JuMP
m = MCPModel()
```

and add variables just as in JuMP:

---

[1] https://github.com/chkwon/Complementarity.jl
[2] http://pages.cs.wisc.edu/~ferris/path.html

11.1. Linear Complementarity Problems (LCP)

```
@variable(m, q1 >= 0)
@variable(m, q2 >= 0)
```

To define the linear function **F**, we use the `@mapping` macro:

```
@mapping(m, F1, - (a - c1 - 2b*q1 - b*q2) )
@mapping(m, F2, - (a - c2 - 2b*q2 - b*q1) )
```

We set up the complementarity condition using the `@complementarity` macro:

```
@complementarity(m, F1, q1)
@complementarity(m, F2, q2)
```

Then solve the problem and show the solution:

```
solveMCP(m)
@show result_value(q1)
@show result_value(q2)
```

The default solver is the PATH Solver and we obtain the following result:

```
Could not open options file: path.opt
Using defaults.
2 row/cols, 4 non-zeros, 100.00% dense.

Path 4.7.03 (Thu Jan 24 15:44:12 2013)
Written by Todd Munson, Steven Dirkse, and Michael Ferris

INITIAL POINT STATISTICS
Maximum of X. . . . . . . . . .   2.6667e+00 var: (q1)
Maximum of F. . . . . . . . . .   2.2204e-16 eqn: (F1)
Maximum of Grad F . . . . . . .   2.0000e+00 eqn: (F1)
                                             var: (q1)

INITIAL JACOBIAN NORM STATISTICS
Maximum Row Norm. . . . . . . .   3.0000e+00 eqn: (F1)
```

```
Minimum Row Norm. . . . . . . .   3.0000e+00 eqn: (F1)
Maximum Column Norm . . . . . .   3.0000e+00 var: (q1)
Minimum Column Norm . . . . . .   3.0000e+00 var: (q1)

Crash Log
major  func  diff  size  residual     step       prox     (label)
  0     0                2.2204e-16              0.0e+00  (F1)

Major Iteration Log
major minor  func  grad  residual     step  type  prox     inorm   (label)
  0     0     1     1   2.2204e-16          I    0.0e+00  2.2e-16  (F1)

FINAL STATISTICS
Inf-Norm of Complementarity . .  2.2204e-16 eqn: (F1)
Inf-Norm of Normal Map. . . . .  2.2204e-16 eqn: (F1)
Inf-Norm of Minimum Map . . . .  4.4409e-16 eqn: (F1)
Inf-Norm of Fischer Function. .  2.2204e-16 eqn: (F1)
Inf-Norm of Grad Fischer Fcn. .  4.4409e-16 eqn: (F1)
Two-Norm of Grad Fischer Fcn. .  4.9651e-16

FINAL POINT STATISTICS
Maximum of X. . . . . . . . . .  2.6667e+00 var: (q1)
Maximum of F. . . . . . . . . .  2.2204e-16 eqn: (F1)
Maximum of Grad F . . . . . . .  2.0000e+00 eqn: (F1)
                                            var: (q1)

 ** EXIT - solution found.

Major Iterations. . . . 0
Minor Iterations. . . . 0
Restarts. . . . . . . . 0
Crash Iterations. . . . 0
Gradient Steps. . . . . 0
Function Evaluations. . 1
Gradient Evaluations. . 1
Basis Time. . . . . . . 0.000005
Total Time. . . . . . . 0.000502
Residual. . . . . . . . 2.220446e-16
:Solved

result_value(q1) = 2.6666666666666665
result_value(q2) = 1.6666666666666667
```

11.1. Linear Complementarity Problems (LCP)

To solve the same problem with `NLsolve`, run this:

```
status = solveMCP(m, solver=:NLsolve)
@show result_value(q1)
@show result_value(q2)
```

The result will look like:

```
Results of Nonlinear Solver Algorithm
 * Algorithm: Trust-region with dogleg and autoscaling
 * Starting Point: [2.66667, 1.66667]
 * Zero: [2.66667, 1.66667]
 * Inf-norm of residuals: 0.000000
 * Iterations: 0
 * Convergence: true
   * |x - x'| < 0.0e+00: false
   * |f(x)| < 1.0e-08: true
 * Function Calls (f): 1
 * Jacobian Calls (df/dx): 1

result_value(q1) = 2.666666666666195
result_value(q2) = 1.6666666666667798
```

For the details of the status output, visit the website[3] of `NLsolve.jl`.

A complete code is presented:

Listing 11.1: A simple LCP example `code/chap11/simple_lcp.jl`

```
a = 10; b = 1; c1 = 3; c2 = 4

using Complementarity, JuMP
m = MCPModel()

@variable(m, q1 >= 0)
@variable(m, q2 >= 0)

@mapping(m, F1, - (a - c1 - 2b*q1 - b*q2) )
```

---
[3]https://github.com/JuliaNLSolvers/NLsolve.jl

Chapter 11. Complementarity Problems

```
@mapping(m, F2, - (a - c2 - 2b*q2 - b*q1) )

@complementarity(m, F1, q1)
@complementarity(m, F2, q2)

solveMCP(m)
@show result_value(q1)
@show result_value(q2)

status = solveMCP(m, solver=:NLsolve)
@show result_value(q1)
@show result_value(q2)
```

Instead of using q1 and q2 separately, we can also use an array form. See the following code for the same problem:

Listing 11.2: A simple LCP example *code/chap11/simple_lcp_array.jl*

```
a = 10; b = 1; c = [3, 4]

using Complementarity, JuMP
m = MCPModel()

@variable(m, q[1:2] >= 0)
@mapping(m, F[i in 1:2], - (a - c[i] - 2b*q[i] - b*q[i%2+1]) )
@complementarity(m, F, q)

solveMCP(m)
@show result_value.(q)

status = solveMCP(m, solver=:NLsolve)
@show result_value.(q)
```

Note that I used i%2+1 to denote the other player's index. The a%b operator gives you the remainder after dividing a by b. For example:

```
julia> 0 % 2
0
```

11.1. Linear Complementarity Problems (LCP)

```
julia> 1 % 2
1

julia> 2 % 2
0

julia> 3 % 2
1
```

When i=1, we have i%2+1=2, and when i=2, we have i%+1=1. Also note that `result_value.(q)` is used with a dot, because q is an array.

If you have solved LCP before, you could be familiar with `transmcp.gms`[4] written in GAMS. It models Dantzig's original transportation model[5] as an LCP with GAMS, which is an optimization modeling language. A Julia translation of the same problem is provided below:

Listing 11.3: Dantzig's transportation model as an LCP
code/chap11/transmcp.jl

```julia
using Complementarity, JuMP

plants = ["seattle", "san-diego"]
markets = ["new-york", "chicago", "topeka"]

capacity = [350, 600]
a = Dict(plants .=> capacity)

demand = [325, 300, 275]
b = Dict(markets .=> demand)

distance = [ 2.5 1.7 1.8 ;
             2.5 1.8 1.4 ]
d = Dict()
for i in 1:length(plants), j in 1:length(markets)
    d[plants[i], markets[j]] = distance[i,j]
end
```

---

[4]http://www.gams.com/modlib/libhtml/transmcp.htm
[5]Dantzig, G B, Chapter 3.3. In Linear Programming and Extensions. Princeton University Press, Princeton, New Jersey, 1963.

```
f = 90

m = MCPModel()
@variable(m, w[i in plants] >= 0)
@variable(m, p[j in markets] >= 0)
@variable(m, x[i in plants, j in markets] >= 0)

@NLexpression(m, c[i in plants, j in markets], f * d[i,j] / 1000)

@mapping(m, profit[i in plants, j in markets], w[i] + c[i,j] - p[j])
@mapping(m, supply[i in plants], a[i] - sum(x[i,j] for j in markets))
@mapping(m, fxdemand[j in markets], sum(x[i,j] for i in plants) - b[j])

@complementarity(m, profit, x)
@complementarity(m, supply, w)
@complementarity(m, fxdemand, p)

status = solveMCP(m)

@show result_value.(x)
@show result_value.(w)
@show result_value.(p)
```

I believe the above Julia code is self-explanatory.

11.2 Nonlinear Complementarity Problems (NCP)

As an NCP example, we consider a traffic equilibrium problem. On a transportation network, we let $\mathcal{W}$ be the set of origin-destination (OD) pairs, $\mathcal{P}_w$ be the set of paths for OD pair $w \in \mathcal{W}$, and T_w be the travel demand for OD pair $w \in \mathcal{W}$. We let $\mathcal{P} = \bigcup_{w \in \mathcal{W}} \mathcal{P}_w$. The travel time function $c_p(\mathbf{h})$ for each path $p \in \mathcal{P}$ is defined a function of the traffic flow vector $\mathbf{h} = (h_p : p \in \mathcal{P})$. We also let u_w be the minimum travel time for OD pair $w \in \mathcal{W}$. Then the following NCP can be formulated (see Chapter 8 of Friesz and Bernstein, 2016)[6]:

$$(c_p - u_w)h_p = 0 \qquad \forall p \in \mathcal{P}_w, w \in \mathcal{W}$$

---

[6] Friesz, T.L. and Bernstein, D., 2016. Foundations of network optimization and games. Springer. Vancouver

11.2. Nonlinear Complementarity Problems (NCP)

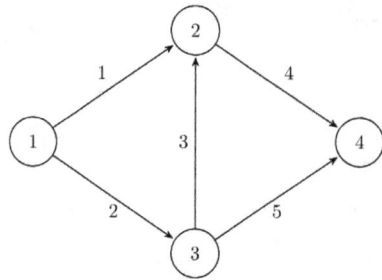

Figure 11.1: A simple network with 4 nodes and 5 arcs.

$$c_p - u_w \geq 0 \qquad \forall p \in \mathcal{P}_w, w \in \mathcal{W}$$
$$h_p \geq 0 \qquad \forall p \in \mathcal{P}_w, w \in \mathcal{W}$$
$$\left(\sum_{p \in \mathcal{P}_\sqsupseteq} h_p - T_w\right) u_w = 0 \qquad \forall w \in \mathcal{W}$$
$$\sum_{p \in \mathcal{P}_\sqsupseteq} h_p - T_w \geq 0 \qquad \forall w \in \mathcal{W}$$
$$u_w \geq 0 \qquad \forall w \in \mathcal{W}$$

Consider a simple example from Section 5.8 of Friesz (2010)[7] with 4 nodes and 5 arcs presented in Figure 11.1. Since there is a single OD pair from node 1 to node 5, we omit the subscript w in this example. The set of nodes and the set of arcs are

$$\mathcal{N} = \{1, 2, 3, 4\} \quad \text{and} \quad \mathcal{A} = \{1, 2, 3, 4, 5\}.$$

respectively. There are three paths available in the path set $\mathcal{P} = p_1, p_2, p_3$, where

$$p_1 = \{1, 4\}, \quad p_2 = \{2, 3, 4\}, \quad p_3 = \{2, 5\}.$$

For each arc a, the travel cost function is defined as follows:

$$c_a = A_a + B_a(f_a)^2 \quad \forall a \in \mathcal{A}.$$

---

[7]Friesz, T.L., 2010. Dynamic optimization and differential games (Vol. 135). Springer Science & Business Media.

where A_a and B_a are positive constants and f_a is the traffic volume at arc a. We define the arc-path incidence matrix:

$$\Delta = (\delta_{ap} : a \in \mathcal{A}, p \in \mathcal{P})$$

where $\delta_{ap} = 1$ if path p contains arc a and $\delta_{ap} = 0$ otherwise. For this example, we have

$$\Delta = \begin{bmatrix} 1 & 0 & 0 \\ 0 & 1 & 1 \\ 0 & 1 & 0 \\ 1 & 1 & 0 \\ 0 & 0 & 1 \end{bmatrix}.$$

Then we can write

$$f_a = \sum_{p \in \mathcal{P}} \delta_{ap} h_p$$

for each $a \in \mathcal{A}$. The travel time function for each path $p \in \mathcal{P}$ can also be written as:

$$\begin{aligned} c_p &= \sum_{a \in \mathcal{A}} \delta_{ap} c_a \\ &= \sum_{a \in \mathcal{A}} \delta_{ap} \left[A_a + B_a (f_a)^2 \right] \\ &= \sum_{a \in \mathcal{A}} \delta_{ap} \left[A_a + B_a \left(\sum_{p' \in \mathcal{P}} \delta_{ap'} h_{p'} \right)^2 \right]. \end{aligned}$$

We assume the travel demand from node 1 to node 5 is $Q = 100$. We use the following data for arc travel time functions:

a	A_a	B_a
1	25	0.010
2	25	0.010
3	75	0.001
4	25	0.010
5	25	0.010

A complete Julia code is provided:

11.2. Nonlinear Complementarity Problems (NCP)

Listing 11.4: Traffic equilibrium problem as an NCP
code/chap11/traffic_equilibrium.jl

```julia
using Complementarity, JuMP

arcs = 1:5
paths = 1:3

delta = [ 1 0 0 ;
          0 1 1 ;
          0 1 0 ;
          1 1 0 ;
          0 0 1 ]
A = [25, 25, 75, 25, 25]
B = [0.01, 0.01, 0.001, 0.01, 0.01]

Q = 100

m = MCPModel()

@variable(m, h[paths] >= 0)
@variable(m, u >= 0)

@mapping(m, excess_cost[p in paths],
    sum(delta[a,p]*(A[a] + B[a]*(sum(delta[a,pp]*h[pp] for pp in paths))^2)
        for a in arcs) - u )
@mapping(m, excess_demand, sum(h[p] for p in paths) - Q)

@complementarity(m, excess_cost, h)
@complementarity(m, excess_demand, u)

status = solveMCP(m)

@show result_value.(h)
@show result_value(u)
```

The result is:

```
result_value.(h) = h: 1 dimensions:
[1] = 49.99999999970811
[2] = 0.0
```

```
[3] = 49.99999999970811

result_value(u) = 99.99999999840261
```

For solving traffic equilibrium problems, one can use the `TrafficAssignment` package[8], which is a native Julia implementation of various Franke-Wolfe methods.

11.3 Mixed Complementarity Problems (MCP)

In the most general form, the Mixed Complementarity Problems have the following structure:
$$\mathbf{F}(\mathbf{z}) \perp \mathbf{l} \leq \mathbf{z} \leq \mathbf{u}$$
which implies
$$\text{if } z_j = l_j, \text{ then } \mathbf{F}_j(\mathbf{z}) \geq 0$$
$$\text{if } l_j < z_j < u_j, \text{ then } \mathbf{F}_j(\mathbf{z}) = 0$$
$$\text{if } z_j = u_j, \text{ then } \mathbf{F}_j(\mathbf{z}) \leq 0$$

Modeling of MCP can be done similarly in the following form:

```
@variable(m, lb[j] <= z[j in 1:n] <= ub[j])
@mapping(m, F[j in 1:n], ...some expression with j...)
@complementarity(m, F, z)
```

where n is the dimension of the problem, and `lb` and `ub` are lower and upper bounds, respectively.

---

[8] https://github.com/chkwon/TrafficAssignment.jl

11.3. Mixed Complementarity Problems (MCP)

12

Parameters in Optimization Solvers

In this chapter, we study how we can set parameters for optimization solvers, in particular CPLEX and Gurobi.

12.1 Setting CPU Time Limit

Mixed Integer Linear Optimization (MILP) problems are popular in practical and research problems. When solvers take particularly long time, one may wish to limit the CPU time. This is useful when we develop a fast heuristic algorithm and want to compare its performance with CPLEX or Gurobi. For example, run CPLEX or Gurobi for 24 hours and compare the solution qualities.
For CPLEX:

```
using JuMP, CPLEX
m = Model(optimizer_with_attributes(
        CPLEX.Optimizer,
        "CPX_PARAM_TILIM" => 86400
     ))
```

For Gurobi:

12.2. Setting the Optimality Gap Tolerance

```
using JuMP, Gurobi
m = Model(optimizer_with_attributes(
        Gurobi.Optimizer,
        "TimeLimit" => 86400
    ))
```

Note that 86400 seconds are 24 hours.

12.2 Setting the Optimality Gap Tolerance

Another way to limit the time spent on solving MILP problems is to set the optimality gap tolerance. The optimality gap is measured by the difference between the best lower and upper bounds. There are two types of differences: absolute gap and relative gap.

$$(\text{absolute gap}) = (\text{upper bound}) - (\text{lower bound})$$

and

$$(\text{relative gap}) = \frac{(\text{upper bound}) - (\text{lower bound})}{(\text{upper bound}) + 10^{-10}}$$

in case of minimization. The constant 10^{-10} is added just to avoid dividing by zero. The solver will stop when the gap becomes less than the tolerance.
For CPLEX:

```
using JuMP, CPLEX

# absolute gap
m = Model(optimizer_with_attributes(
        CPLEX.Optimizer,
        "CPX_PARAM_EPAGAP" => 1e-1
    ))

# relative gap
m = Model(optimizer_with_attributes(
        CPLEX.Optimizer,
        "CPX_PARAM_EPGAP" => 1e-2
    ))
```

For Gurobi:

```
using JuMP, Gurobi

# absolute gap
m = Model(optimizer_with_attributes(
          Gurobi.Optimizer,
          "MIPGapAbs" => 1e-1
        ))

# relative gap
m = Model(optimizer_with_attributes(
          Gurobi.Optimizer,
          "MIPGap" => 1e-2
        ))
```

One may also combine the CPU time limit with the optimality gap tolerance as follows:

```
using JuMP, CPLEX
m = Model(optimizer_with_attributes(
          CPLEX.Optimizer,
          "CPX_PARAM_TILIM" => 86400,
          "CPX_PARAM_EPGAP" => 1e-2
        ))
```

which terminates the solver after 24 hours of computing or when the relative gap is less than 1%.

12.3 Warmstart

When solving MILP problems, we can provide initial solutions to start with. If the quality of the initial solutions are good, then we may be able to reduce the computational time significantly. This is useful when we have relatively good solutions on hand either produced by a heuristic algorithm, or generated in a previous iteration.

We use the `set_start_value` function:

```
using JuMP, Gurobi
m = Model(Gurobi.Optimizer)
@variable(m, x[1:10] >= 0)

init_x = [3, 4, 2, 5, 3, 2, 1, 6, 8, 1]
for i in 1:10
  set_start_value(x[i], init_x[i])
end

# Or simply:
set_start_value.(x, init_x)
```

We don't need to supply initial solutions to all variables. We can provide initial solutions for some variables only; optimization solvers will guess other solutions by themselves. Well, although solvers can guess unspecified initial values, it is always almost best to "guess" by yourself. If you know of only part of a good solution, try first to find the entire solution and provide it to the solver.

12.4 Big-M and Integrality Tolerance

Big-M has been popularly used for linearization of nonlinear terms—especially bilinear terms—in mixed integer optimization problems. While techniques with big-M are practically useful, we need to address an important question: How big is big and how big should it be?

Suppose x is a binary variable and y is a nonnegative continuous variable. Consider a bilinear term xy, which may be linearized with an additional variable $w = xy$ as follows:

$$w \geq 0$$
$$w \leq Mx$$
$$w \leq y$$
$$w \geq y + M(x - 1)$$

Let's check if the above is correct. When $x = 0$, we have:

$$w \geq 0$$

$$w \leq 0$$
$$w \leq y$$
$$w \geq y - M$$

The above four inequalities guarantee that $w = 0$ as long as $y - M \leq 0$. When $x = 1$, we have:

$$w \geq 0$$
$$w \leq M$$
$$w \leq y$$
$$w \geq y$$

In this case, we have $w = y$ if, again, $M \geq y$.

One "little" problem here is that we do not know the value of y in advance. Often there are two choices:

1. We find an upper bound on y, and set M as the upper bound.

2. We set M as an arbitrarily big number.

If the first approach, called 'bounding big M', is possible, it's the best. Try to find the smallest upper bound. When we use the second approach, we simply don't know how big it should be; hence often come up with a very big number.

What's wrong with a very big number? Well, a critical problem exists with the term $M(1 - x)$. Optimization solvers such as Gurobi and CPLEX find binary solutions with some numerical computation and think that a number is 1 if it is close enough to 1.0. That means, 0.999999984 can be 1 within optimization solvers. Then, if M is as big as 1000000000, then the term $M(1 - x)$ becomes 16, while it should be zero. Well, in my Julia, it is not even 16.

```
julia> 1000000000*(1-0.999999984)
15.999999991578306
```

One remedy for this issue can be adjusting the integrality tolerance of optimization solvers. Both in CPLEX and Gurobi, it is 10^{-5} by default. One may wish to set a smaller value for it; for example 10^{-9}. For CPLEX:

12.5. Turning off the Solver Output

```
using CPLEX
m = Model(optimizer_with_attributes(CPLEX.Optimizer, "CPX_PARAM_EPINT" => 1e-9 ))
```

For Gurobi:

```
using Gurobi
m = Model(optimizer_with_attributes(Gurobi.Optimizer, "IntFeasTol" => 1e-9 ))
```

With a smaller integrality tolerance, the computation will obviously be slower. Also note that the integrality tolerance issue is not the only problem that big M creates. So, try to make big M small—but not too small.

12.5 Turning off the Solver Output

CPLEX and Gurobi by default display many useful outputs while solving optimization problems, mostly useful but sometimes unnecessary. When you would like to turn output displays off, use the following codes:

```
using CPLEX
m = Model(optimizer_with_attributes(CPLEX.Optimizer, "CPX_PARAM_SCRIND" => 0 ))
```

For Gurobi:

```
using Gurobi
m = Model(optimizer_with_attributes(Gurobi.Optimizer, "OutputFlag" => 0 ))
```

12.6 Other Solver Parameters

In addition to the CPU time limit, integrality tolerance, and solver output options, there are many other parameter values that we can control. For the full lists of parameters, visit the following links:

Chapter 12. Parameters in Optimization Solvers

- CPLEX: List of CPLEX parameters[1] — Julia uses parameter names for C.
- Gurobi: Parameters[2]

---

[1] Visit https://www.ibm.com/support/knowledgecenter and search by "List of CPLEX parameters." You can also browse this PDF file.

[2] http://www.gurobi.com/documentation/8.1/refman/parameters.html

Index

Symbols
.nl format 171

A
add_edge! 142
adjacency matrix 187
affine scaling algorithm 151
AMPL 171
AmplNLWriter 172
argmin 113, 211
Array 46, 109, 181, 212
array 41
    tuples 47, 128
@assert 99, 204
automatic differentiation 91

B
Bernoulli distribution 178
bi-level optimization 171
big-M 242
Bonmin 171

box-constrained optimization 168
break 55
Broyden-Fletcher-Goldfarb-Shanno,
       BFGS 168

C
Calculus 86
Clp 104
COIN-OR 30, 172
collect 50, 100, 110
collection 54
colon (:) 50
combinations 100, 110
Combinatorics 100
comma separated values, CSV 68,
       125, 127, 135
commodity 123
Complementarity 227
    @complementarity 228
    @mapping 228

247

complementarity
    linear complementarity problems, LCP 225
    mixed complementarity problems, MCP 237
    nonlinear complementarity problems, NCP 233
complementarity problems 225
conjugate gradient 168
`convert` 203
`copy` 190
Couenne 171
CPLEX 3, 104, 239
    installation 16
    parameters 245
CPU time 239
curve fitting 79
`curve_fit` 82

D
DelimitedFiles 70
`derivative` 86, 93
`Diagonal` 154
`Dict` 55, 128, 136, 146
dictionary 54
Dijkstra's algorithm 140, 144
`dijkstra_shortest_paths` 142
Distributions 177
`dot` 43, 103

E
`end` 49
`enumerate_paths` 142
equilibrium
    traffic equilibrium 233
    Wardrop equilibrium 233

F
`fieldnames` 108
file input/output 67
`findall` 113, 115, 191
`findfirst` 113
finite difference 84
`Fminbox` 169
`for` 54
ForwardDiff 92
`function` 57

G
GAMS 3
Gauss-Kronrod integration method 90
Geometric distribution 178
global optimization 171
GLPK 104
Golden Section 167
`gradient` 87, 93
graph 123
Gurobi 4, 104, 239
    installation 15
    parameters 245

H
`hessian` 87, 93

I
identity matrix 43
IJulia 18
importance sampling 195
index 48
`Inf` 99, 126
inner product 43
integrality tolerance 242
`inv` 44, 101, 154

inverse matrix 44
Ipopt 169
`!isempty` 147

J
Julia
    installation 4
        macOS 10
        Windows 5
JuMP 3
    `Bin` 39
    `@constraint` 32, 35, 38
    `getobjectivevalue` 130
    `Int` 39
    `@NLconstraint` 170
    `@NLobjective` 170
    `@objective` 32, 35
    `optimize` 33
    `print()` 32
    `set_start_value` 241
    `shadow_price()` 33
    `value` 33
    `@variable` 32, 35

L
`label` 156
Lagrangian relaxation 197
least-squares fit 81
`legend` 156
`length` 53, 55, 127, 204
Levenberg-Marquardt algorithm . . . 81
LightGraphs 141
line search 165
linear programming, LP 30, 95
linear regression 80
LinearAlgebra 43, 99, 154

linearization 242
link 123
loop 54
lower bound 199
LsqFit 81

M
`margin_error` 82
mathematical program with
        complementarity conditions,
        MPCC 172
MATLAB 3
matplotlib 72
matrix 41
`max` 127
maximum 127, 211
minimal-cost network-flow problem
        123
mixed integer linear programming,
        MILP 38
mixed integer nonlinear programming,
        MINLP 171, 175
`module` 116
Monte Carlo 177

N
Nelder-Mead 168
network optimization 123
NLsolve 227
node 123
nonconvex nonlinear optimization 171
nonlinear optimization 169
nonlinear programming, NLP 169
`norm` 154
Normal distribution 64, 177
    multi-variate 179

`normcdf` 66
`norminvcdf` 66
`normpdf` 66
numerical differentiation 84
numerical integration 87
numerical methods 79

O

`ones` 44, 214
Optim 165, 168
optimality gap 199, 240
optimization 165
    bi-level 171
    box-constrained 168
    global 171
    nonconvex nonlinear 171
    nonlinear 169
    unconstrained 165
`optimize` 167, 168

P

p-median problem 201
package management 21
packages
    AmplNLWriter 172
    Calculus 86
    Combinatorics 99, 100
    Complementarity 227
    CPLEX 17
    DelimitedFiles 70
    Distributions 177
    ForwardDiff 92
    GLPK 7, 12
    Gurobi 16
    IJulia 19
    Ipopt 169
    LightGraphs 141
    LinearAlgebra 43, 99, 154
    LsqFit 81
    NLsolve 227
    Optim 165, 168
    PathDistribution 195
    PATHSolver 227
    Plots 72
    Printf 53
    PyPlot 72, 156, 218
    QuadGK 90
    StatsFuns 65
PathDistribution 195
PATHSolver 227
`plot` 156
Plots 72
plotting 72
primal path following algorithm . . 157
`print` 52
Printf 53
`@printf` 53
`println` 52
probability distribution
    Uniform distribution 63
probability distributions 177
    Bernoulli 178
    Binomial 66
    Gamma 66
    Geometric 178
    Normal 66, 177
        multi-variate 179
    Normal distribution 64
    Poisson 66
`push!` 214
PyPlot 72, 156, 218

Python 3

Q
QuadGK 90
quadgk 90

R
rand 63, 178, 192
randn 64
random number 63
randomized linear program 179
range 48
rank 99
readdlm 70
REPL 21
revenue management 179
Riemann sum 88
round 126

S
scope blocks 60
scope of variables 59
second_derivative 87
semicolon (;) 42
Set 146
setdiff 110, 115, 148
shortest path problem 139
simplex method 95
Simpson's rule 90
simulated annealing 168
sink node 123
size 99, 109, 204
solvers 29
    Bonmin 172
    Cbc 29
    Clp 29
    Couenne 172
    CPLEX 16, 29
    GLPK 7, 12, 29
    Gurobi 15, 29
    parameters 239, 245
sorting 207
sortperm 207
source node 123
@sprintf 54
StatsFuns 65
subgradient optimization 200
sum 182, 214

T
Taylor series 85
tolerance
    integrality 242
    optimality gap 240
traffic assignment 233
traffic equilibrium 233
transportation problem 133
transpose 43, 154, 214
trapezoidal rule 89
Tuple 128, 146
tuple 47
type 107
typeof 108

U
unconstrained optimization 165
undef 46
upper bound 199

V
vector 41

W

Wardrop equilibrium 233
warmstart 241
`while` 57, 191

Z

`zeros` 44, 99
    `Int` 208

www.ingramcontent.com/pod-product-compliance
Lightning Source LLC
Chambersburg PA
CBHW081554220526
45468CB00010B/2658